Artificial Neural Networks - Recent Advances, New Perspectives and Applications

Edited by Patrick Chi-Leung Hui

Published in London, United Kingdom

Artificial Neural Networks - Recent Advances, New Perspectives and Applications
http://dx.doi.org/10.5772/intechopen.100662
Edited by Patrick Chi-Leung Hui

Contributors
Vidyadevi G. Biradar, Piyush Kumar Pareek, S.G. Mohan, H.C Nagaraj, Mustafa Turker, Alio Boubacar Goga, Bin Du, Peter D. Lund, Eugenio Fazio, Alessandro Bile, Hamed Tari, Mario Acosta Flores, Marta Lilia Eraña-Díaz, Hamza Abubakar, Muntari Idris, Shafaati Akbar, Pourazad Hamidreza

First published in London, United Kingdom, 2023 by IntechOpen
IntechOpen is the global imprint of INTECHOPEN LIMITED, registered in England and Wales, registration number: 11086078, 5 Princes Gate Court, London, SW7 2QJ, United Kingdom

British Library Cataloguing-in-Publication Data
A catalogue record for this book is available from the British Library

Additional hard and PDF copies can be obtained from orders@intechopen.com

Artificial Neural Networks - Recent Advances, New Perspectives and Applications
Edited by Patrick Chi-Leung Hui
p. cm.

This title is part of the Artificial Intelligence Book Series, Volume 13
Topic: Computational Neuroscience
Series Editor: Andries Engelbrecht
Topic Editor: Magnus Johnsson

Print ISBN 978-1-83769-994-0
Online ISBN 978-1-83768-222-5
eBook (PDF) ISBN 978-1-83768-223-2
ISSN 2633-1403

IntechOpen Book Series

Artificial Intelligence

Volume 13

Aims and Scope of the Series

Artificial Intelligence (AI) is a rapidly developing multidisciplinary research area that aims to solve increasingly complex problems. In today's highly integrated world, AI promises to become a robust and powerful means for obtaining solutions to previously unsolvable problems. This Series is intended for researchers and students alike interested in this fascinating field and its many applications.

Meet the Series Editor

Andries Engelbrecht received the Masters and Ph.D. degrees in Computer Science from the University of Stellenbosch, South Africa, in 1994 and 1999 respectively. He is currently appointed as the Voigt Chair in Data Science in the Department of Industrial Engineering, with a joint appointment as Professor in the Computer Science Division, Stellenbosch University. Prior to his appointment at Stellenbosch University, he has been at the University of Pretoria, Department of Computer Science (1998-2018), where he was appointed as South Africa Research Chair in Artifical Intelligence (2007-2018), the head of the Department of Computer Science (2008-2017), and Director of the Institute for Big Data and Data Science (2017-2018). In addition to a number of research articles, he has written two books, Computational Intelligence: An Introduction and Fundamentals of Computational Swarm Intelligence.

Meet the Volume Editor

Patrick Chi-Leung Hui obtained an MSc in Technological Economics from the University of Stirling, UK in 1988; an MSc in Information Systems from Hong Kong Polytechnic University in 1992; an MSc (Eng) in Computers in Manufacturing from the University of Hong Kong in 1995; a Ph.D. from Hong Kong Polytechnic University in 1999; an LLB (Hons) from the University of Wolverhampton, UK in 2004; and an LLM in Information Technology and Intellectual Property Law from the University of Hong Kong in 2008. He is a chartered engineer with professional diplomas in marketing, finance, and accounting. He is also a chartered member of the British Computer Society and the Chartered Institute of Marketing. He has published more than fifty journal and conference papers. Dr. Hui is a project fellow in the Department of Computing, at Hong Kong Polytechnic University.

Contents

Preface

Although artificial neural networks (ANNs) have been around for decades, they have only begun to be used in fields such as health care, energy management, supply chains, inventory management, and transportation due to unprecedented advances in information and communication technologies.

This book explores the potential of ANNs for applications in different fields. It includes eight chapters that discuss deep learning, ANN tools, and other cutting-edge technologies. It also suggests avenues for further research into ANN techniques for medical imaging to detect breast tumors, classification of COVID-19 surveillance datasets, health management, estimation of materials processing parameters, solar energy management, and control of a petrochemical unit.

Patrick Chi-Leung Hui
Department of Computing,
Hong Kong Polytechnic University,
Kowloon, Hong Kong

Chapter 1

Artificial Intelligence at the Service of Medical Imaging in the Detection of Breast Tumors

Alio Boubacar Goga

Abstract

Artificial intelligence is currently capable of imitating clinical reasoning in order to make a diagnosis, in particular that of breast cancer. This is possible, thanks to the exponential increase in medical images. Indeed, artificial intelligence systems are used to assist doctors and not replace them. Breast cancer is a cancerous tumor that can invade and destroy nearby tissue. Therefore, early and reliable detection of this disease is a great asset for the medical field. Some people use medical imaging techniques to diagnose this disease. Given the drawbacks of these techniques, diagnostic errors of doctors related to fatigue or inexperience, this work consists of showing how artificial intelligence methods, in particular artificial neural networks (ANN), deep learning (DL), support vector machines (SVM), expert systems, fuzzy logic can be applied on breast imaging, with the aim of improving the detection of this global scourge. Finally, the proposed system is composed of two (2) essential steps: the tumor detection phase and the diagnostic phase allowing the latter to decide whether the tumor is benign or malignant.

Keywords: breast cancer, artificial intelligence, artificial neural network, deep learning, expert system, fuzzy logic, medical imaging, big data

1. Introduction

Breast cancer is a disease in which cells in breast tissue change and divide in an uncontrolled manner, usually producing a lump or lump. Most breast cancers start in the lobules (mammary glands) or in the ducts that connect the lobules to the nipple. If not diagnosed early, it can lead to death. It can be divided into two (2) groups: normal and abnormal and it can also be divided into two (2) categories: benign (not dangerous) and cancerous (malignant). Benign tumors grow quite slowly and do not invade neighboring tissues or spread to different parts of the body [1]. The early and reliable detection of it focuses on reviewing data from past diagnoses and gathering valuable information from past data. Currently, the early detection and diagnosis of tumors using image processing techniques and artificial learning can be of great help in improving the accuracy of a breast cancer diagnosis. Secondly, medical imaging plays a major role in the clinical diagnosis of diseases, the evaluation of treatment and the detection of abnormalities in different organs of the body such as [2, 3]. In addition, several researchers have focused intensively on the production and interpretation of medical images to identify the majority of diseases including breast cancer. These images thus facilitate the identification of the disease and help in the detection of a pathological lesion, in the clinical treatment of the patient. Artificial intelligence has played a major role in the medical field, such as the analysis of medical images. It is the most effective way to detect breast cancer, with regular use of different modalities such as MRI, mammography, computed tomography and radiographic ultrasound. The most frequently used images are mammography, ultrasound, MRI, histology and thermography [4]. Mammography can detect and diagnose breast cancer in women. Mammography images can be examined by professional radiologists to determine if there are any abnormalities in the breast. She may show breast changes for up to a year or two before the patient or the doctor sees the symptoms. The American Cancer Society (2019–2020) recommends a mammogram once a year for all women over 40. Dense breast tissue during a mammogram may appear white or light gray. This may make it easier to view mammograms in younger women who appear to have thicker breasts. Therefore, it is ineffective in patients under 40 years of age, with dense breasts and less sensitive to small tumors. Most breast diseases look like signs of cancer and require tests to identify them, and often a biopsy [5]. Another method of breast cancer screening is ultrasound imaging which can be used to supplement mammography by determining the liquid or solid nature of a lesion, especially in women with large breasts [6]. Magnetic resonance imaging (MRI) is another technique for early detection of cancer cells, in addition to ultrasound and mammography techniques [7]. Despite rapid advances in medical research, the benchmark for cancer diagnosis remains histopathological diagnosis [8]. Another breast cancer imaging modality is thermography or thermal imaging of the breast which is a painless and non-invasive method that is often used to detect changes in the breast that may indicate this global scourge [9]. Finally, the use of artificial intelligence makes it possible to identify candidate biomarkers for medical imaging [10].

2. Artificial intelligence and medical imaging

Artificial intelligence is at the crossroads of several fields. Among these fields we can cite computer science, mathematics, medicine, physics, philosophy, etc. Early detection plays a very important role in the diagnosis of cancer, especially the diagnosis of breast cancer. It can promote the chances of recovery from it, therefore it is able to improve the long-term survival rates of patients. Note that, medical imaging has long been used to perform early detection of breast cancer, its monitoring and post-treatment follow-up, nevertheless, the direct interpretation of a large number of medical images is a difficult task. and depends on the expertise of the radiologist. In fact, the interpretation of medical images still relies today on the eye of the radiologist or the doctor. Several imperfections mar this process. The human eye is fallible, fatigable, subject to many cognitive biases, and its performance depends on its experience. In addition, its relevance depends on the visibility (salience) of the images to be located. A large lesion, of which the contrast is high will be easily detected, this is not the case when the lesion is small and of low contrast and if it is located outside the nosological field questioned. To solve this problem, ordinary assisted diagnostic systems were developed as early as the early 1980s. Initially, from the 1970s to the 1990s, medical image analysis was performed from sequential treatments from the treatments of. low level (denoising, contrast enhancement, detection of edges and lines, segmentation of regions) up to pattern recognition through mathematical modeling. Then, to develop automatic analysis systems, the researchers drew inspiration from the human brain to build expert systems, which use artificial intelligence techniques [11, 12].

AI applied to medicine aims to:

- automate the detection of pathological images;
- treat large cohorts of patients;
- allow the detection of incidental lesions, not sought "a priori";
- make the interpretation of images more reliable;
- identify patterns, allowing the classification of lesions;
- establish standardized reports.

Finally, "supervised" artificial intelligence requires a large amount of data allowing the learning of AI methods. The French society of Radiology and the National Federation of Radiological Doctors have decided to create an "ecosystem" of 500 million imaging files thus making a medical imaging database available to researchers [13, 14]. 400 million would be truly exploitable for the development of AI algorithms. This database should also be continuously updated. The potential French database has many advantages and is recognized internationally [15].

3. Artificial intelligence and big data

The exponential increase or the quantitative explosion of data has forced researchers in data science, then in medical science, to transform the way they see and analyze the world. In medicine, this increase is caused by the number of medical images produced. Thus, for mammographic examinations, two (2) or four (4) mammograms are performed per patient and this at the rate of one or two mammograms per breast. Therefore, a woman can have at least two (2) medical images in this context. In this case, it is about understanding new ways to capture, search, share, store, analyze and present data whose order of magnitude grows exponentially. These large-scale data (Big Data) are generally analyzed using artificial intelligence methods. Note that these two (2) concepts are increasingly applied in medical research [16]. There are several sources of medical " big data ", we can cite clinical data from databases such as health insurance, private mutuals or the pooling of cohort data, or even digital traces (keywords typed into an internet server); but also data from medical imaging (a single imaging test that may contain millions of pixels), or even biological data. So, for better data exploitation, Big Data is analyzed by methods derived from AI and its sub-specialty, artificial learning (Machine Learning) [17, 18].

4. Artificial intelligence and data mining

The ability to make good use of data, particularly that of medical imaging, is at the heart of the challenges of tomorrow's medicine. We can cite the development of diagnostic aid tools, radiomics which consists in extracting quantitative data in order to identify potential imaging biomarkers, the development of in-silico models making it possible to accelerate medical research, formulation and validation of hypotheses from the retrospective use of several independent cohorts, patient screening to better target patients eligible for a clinical trial [19]. Indeed, data exists in abundance, nevertheless the exploitation of this "big data" is a very difficult task for doctors and other actors in their specific fields.

5. Artificial intelligence and cognitive psychology

It is certain that the observation, modeling, understanding of cognitive activity and intelligence are, as natural sciences, the responsibility of cognitive psychology, or more generally of cognitive ethology. According to Margaret Bogden artificial intelligence is the art of simulating intelligence using a computer, this clearly falls under the science of the artificial. Insofar as it draws its inspiration from cognitive psychology and cares about psychological realism, it can be an experimental counterpoint all the richer for cognitive psychology (or cognitive ethology) as experimentation using software does not pose the ethical problems posed by human or even animal experimentation. Indeed, cognitive psychology and artificial intelligence present themselves as sister disciplines. AI will have two sides [20]:

- predominantly psychological, it is above all concerned with the realism of simulations of the functioning of the human mind;
- redominantly computer science, it seeks intelligent global behavior, human or not; In addition to purely practical reasons, we can also consider that the human mind probably does not have a monopoly on intelligence, and would benefit from being helped by other forms of intelligence.

6. Artificial intelligence in computer vision

Computer vision is an AI technology. There are interactions between artificial intelligence and computer vision, from the point of view of knowledge-based systems for the interpretation of images and scenes, and for the recognition of shapes, structures or objects in images. The general objective of these approaches is to add semantics to images, by associating visual information extracted from images on the one hand and knowledge or models on the other hand [21].

7. The use of artificial intelligence in clinical practice

The analysis of current medical imaging applications using artificial intelligence for current clinical use provides information on the directions of scientific research to be considered in this field. The first half of 2018 was marked by the arrival of three

(3) AI players in imagery. Their solutions have all been approved, for the first time, by the United States Food and Drug Administration (FDA). This is especially the application Viz.ai (San Francisco, CA, USA) for acute stroke, using the deep learning (Deep Learning) for automatic detection of occlusion of cerebral vessels to the angio-CT and the immediate call of the interventional radiologist on call. This is also the case with the IDx-DR software capable of detecting diabetic retinopathy on the fundus without even the intervention of an ophthalmologist. This app can be used in theory by paramedics for early detection action. According to these authors, its reliability is high. It relates to a clinical study (NCT02963441) of eye funds on 10 American centers with 900 diabetic patients in whom, in 90\% of cases, the IDx-DR solution (Coralville, Iowa, USA) allowed the correct diagnosis [22]. Finally, Osteodetect from the company Imagen (New York, USA) is another tool which makes it possible to accelerate the detection of wrist fractures on standard 2D digitized radiologies. This assisted detection uses artificial intelligence techniques to enable faster diagnosis based on an initial retrospective analysis of 1000 images per second at 24 centers. The device has been announced by its developer to be dedicated above all to non-radiologist nursing staff (general practitioners, emergency physicians, resuscitators, etc.). The site emphasizes that it is a complementary tool and not a software to replace the expertise of radiologists. In [23], these three (3) examples confirm the importance of AI methods in clinical practice.

8. Role of artificial intelligence in medical imaging

The aim of artificial intelligence systems is not to replace radiologists but rather, to provide them with convincing help. So, let the practicing doctors handle the use of artificial intelligence in their specialty [24]. The fields of application of AI techniques in medical imaging are numerous: the creation of examination protocols, improvement of image quality and reduction of the irradiation dose, reduction of acquisition times in MRI, optimization of programming, presentation of images for interpretation, development of detection assistance tools, post-processing assistance, quantification tools, segmentation, image registration, analysis the quality of the images produced, realized production [25].

9. Some definitions

Here are some definitions of artificial intelligence (AI) techniques [26–29]:

- Artificial Neural Networks (ANNs): These are AI techniques aimed at simulating the functioning of neural cells to mimic the functioning of the human brain. They are mainly used in the recognition of speech and images. These techniques can be simulated in software or with specialized electronic circuits.

- Deep learning (DL): it is an extension of artificial learning integrating supervised learning and self-learning functions based on complex and multidimensional data representation models. It is an evolution of the ANNs which have multiple layers and sub-layers of neurons.

- The support vector machine (SVM): it is an algorithm which will classify data according to a linear threshold and whose objective is to solve the problems of classification or discrimination in two classes. Note that there is a modification of this algorithm which allows it to be used for the regression.

- Expert systems: these are AI systems based on high-level knowledge modeling with predicate logics (if this then that, if the patient has her symptoms then the patient has breast cancer, etc.) and rule engines.

- Fuzzy logic: it is an AI technique created by Lofti Zadeh in the 1960s and representing information not in binary form but in fuzzy form between 0 and 1. It is sometimes used in rules engines of expert systems.

10. The application of artificial neural networks

This artificial intelligence technique is used to detect breast cancer. In [30], breast cancer is detected using two electronic noses (EN) to analyze breath and urine samples. Exhaled breath samples were taken from 48 breast cancer patients and 45 healthy patients who served as a control group while urine samples were taken from 37 breast cancer patients. Breast based on mammographic tests and 36 healthy patients. These two ENs made it possible to analyze exhaled respiration on the one hand, and on the other hand the authors used gas chromatography mass spectrometry (GC-MC) to analyze the substances present in the urine. The first EN used was the MK4 model. The second EN used was Cyranose 320. Indeed, the model obtained, that is to say the artificial neural network on the basis of the analysis carried out by the MK4 and Cyranose 320, made it possible to classify the patients suffering from dystrophy. 'breast cancer with an accuracy, on average, of more than 95%.

In [31], the proposed method is based on the representation of images using discrete Haar wavelets. Then, they are introduced into artificial neural networks. These digital images are obtained by biopsy from the Near East University Hospital. The images are classified using two classifiers including the Backward Propagation Neural Network (BPNN) and the Radial Basis Function Network (RBFN).

In [32], to help radiologists quickly detect breast cancer, these authors proposed a computational model based on an artificial neural network. This is able to detect the presence or absence of an abnormality on a mammogram. In order to train or train their model, they used a database of digital mammograms generated by MIAS (Mammographic Image Analysis Society). Then, they used 60 images divided into 30 normal images and 30 images including anomalies. This has been confirmed by expert radiologists. The artificial neural network model created in this study has the following advantages: simplicity of extracting the descriptive parameters of each mammogram, automatic and rapid detection of the presence or not of an anomaly on a mammogram, possibility of adapting the template to other images from different medical bases with different resolutions. Finally, they demonstrated the performance of the model obtained for the detection of breast cancer on a mammogram with a correct recognition rate of 91.66%.

11. The application of deep learning

Deep learning is another form of artificial neural networks. It is the most widely used artificial intelligence method when it comes to the classification of breast cancer on medical images. In [5], in order to help medical experts quickly diagnose breast cancer, the authors presented the Convolutional Neural Network Improvement Algorithm for Breast Cancer Classification (CNNI-BCC). Indeed, the sensitivity of the convolutional neural network (CNN) to radiological images prompted the authors to improve CNN. To detect and categorize into malignant, benign, and normal, the CNNI-BCC method uses data extension by functionality (FWDA) algorithms, convolutional neural network-based classification (CNNBS), and lesion locator based on interactive detection (IDBLL). This model can be incorporated on portable devices such as smartphones. The materials used in this work are digital mammography databases. These bases were prepared and supplied by MIAS. Then the experiments are applied to 21 mild cases, 17 malignant cases and 183 normal cases. CNNI-BCC has achieved an accuracy of 90.50%.

In [33], to alleviate the lack of early detection of breast cancer, the authors proposed a cancer detection approach based on a convolutional neural network (CNNs). This technique can simultaneously locate and classify the mass as benign or malignant on a mammogram image. Then, to train or train, validate and test the method, datasets were collected at various sites, in particular at St Gebriel Hospital, Grum Hospital, Betezatha Hospital, Korean Hospital, Kadisco Hospital and at Pioneer Diagnostic. Indeed, the mammogram images were collected with their document reports that show the results of screening and diagnosis of the patients. Overall, the proposed approach includes the following steps:

- Data collection in different hospitals in Ethiopia,
- Preprocessing mammographic images to improve data quality and prepare them appropriately for deep learning,
- After the preprocessing, the noise on the images is eliminated by applying Gaussian filtering, median filtering and bilateral filtering,
- And later images were enhanced using Adaptive Contrast Limited Histogram Equalization (CLAHE),
- Finally, a morphological operation is performed to extract the breast region from the background and to remove part of the mammographic image such as artifacts, labels, patient profiles and the like.

Ultimately, the model was trained and evaluated via mammographic images and achieved an accuracy of 91.86%.

In [34], in order to help radiologists more precisely diagnose breast cancer, this research proposes the development and validation of a new scheme called SD-CNN (Shallow-Deep Convolutional Neural). This method combines image processing and machine learning techniques to improve diagnosis using full field digital mammography (FFDM) by leveraging information available from contrast enhanced digital mammography (CEDM). The first hypothesis posed by the authors is that the application of a deep CNN (Deep-CNN) to CEDM is able to take advantage of recombinant imaging to improve the classification of breast lesions. Second, with the aim of extending the advantages of the CEDM imaging modality to the FFDM imaging modality, they hypothesized that a shallow CNN (Shallow-CNN) is capable of uncovering non-mapping. Linear between the LE images that is to say the low energy (LE) and recombined images. The objective of this study is to validate these two hypotheses using a single study procedure and two separate data sets, including a data set acquired from a tertiary medical center (Mayo Clinic Ari-zona) and a set public data file from INbreast. They first developed a CEDM Shallow-CNN to discover the relationships between LE images and recombinant images. This Shallow-CNN is then applied to FFDM to restore "virtual" recombined images. In collaboration with FFDM, a trained Deep-CNN is introduced for feature extraction, followed by classification models for diagnosis. The approach proposed by the authors had an accuracy of 90%.

In [35], manual segmentation is time consuming and does not take into account the appearance of anatomical structures. So to solve this problem the authors proposed a method of auto-segmentation of the clinical target volume (CTV) called deeply dilated residual network (DD-ResNet). It performs automatic segmentation in order to plan the computed tomography or scanner. They used data from early-stage breast cancer patients who only underwent breast-conserving therapy from January 2013 to December 2016 at the Radiation Oncology Department of the Cancer Hospital of the Chinese Academy of Medical Sciences. To evaluate their method, they performed a comparison between self-segmentation and manual segmentation. This comparison is based on images from different patients and also of different sizes. The results show that the self-segmented contours of the CTV were close to the manually segmented contours in shape, volume and location. The deep learning algorithm (DD-ResNet) proposed by the authors, could be used to improve consistency in bypassing and streamlining breast cancer radiotherapy processes.

12. The application of support vector machines

This method is also useful in the detection of breast cancer. In [36], the authors propose a wide-margin separation technique to perform the stain classification in the context of breast cancer detection. This method is called, the Support Vector Machine (SVM). In this study, they also dealt with character extraction using the Hough transform. The latter makes it possible to detect the characteristics of the mammographic image in order to provide the values to the classifier, that is to say SVM. The mammography images used are collected from the Mammography Image Analysis Database Company (MIAS). Among the 322 images of this company, 95 images were taken to carry out this work. Note that SVM has a success rate of 94%.

In [37], the use of machine learning algorithms such as Support Vector Machine (SVM), decision trees (C4.5), Naive Bayes (NB) and K nearest neighbors (K-NN for K-Nearest Neighbor in English) in medical sciences can classify and predict breast cancer. The authors compared the performance of these algorithms using Wisconsin breast cancer datasets from the UCI machine learning repository. They managed to prove that SVM has the highest accuracy (97.13%) and the lowest error rate (2%).

13. The application of expert systems

This technique is useful in the diagnosis of breast cancer. In [38], the need for a powerful diagnostic tool motivated the authors to create an expert system for breast cancer diagnosis called Ex-DBC to effectively diagnose breast cancer. To perform the diagnosis, the system uses fuzzy rules. In this study, the mammography mass dataset is provided by the UCI Machine Learning Repository. This dataset can be used to predict the severity (benign or malignant) of a mammographic mass lesion from the attributes of the Breast Imaging Recording And Data System (BI-RADS) and the patient's age. It contains a BIRADS assessment, the patient's age and the three BI-RADS attributes as well as the ground truth (the gravity field) for 516 benign masses and 445 malignant masses that were identified on mammographic images collected at the radiology institute of Erlangen-Nuremberg University between 2003 and 2006. Note that the expert system Ex-DBC has a powerful inference engine containing fuzzy rules that can detect hidden relationships in the unrecognized case by the human expert. The goal of the Ex-DBC is to minimize human error by capturing and interpreting points that may not be recognized by the radiologist. Ultimately, the expert system created in this study can make an important contribution to the prevention of unnecessary biopsy in the diagnosis of breast cancer and it can also be useful in training medical students.

14. The application of fuzzy logic

This technique can detect and diagnose breast cancer. In [39], it is difficult to improve the image and remove noise at the s-ame time. This prompted the authors to propose a new contrast enhancement algorithm based on fuzzy logic and fuzzy entropy. The principle of maximum fuzzy entropy is used to map the original image, then the characteristics of the American image are taken into account. More precisely, the edge and texture information is extracted to evaluate the characteristics of the lesions and the phenomenon of diffusion of the American images and the local information is used to define the enhancement criterion. The algorithm improves the details and characteristics of lesions using local fuzzy information. The proposed method includes the following steps: image normalization, image fuzzification, edge information extraction, texture information extraction and contrast enhancement. Indeed, the images of the American breasts used in this study were provided by the Second Affiliated Hospital of Harbin Medical University (HMU). The database included a total of 86 images from 49 cases, and each unique lesion is in an image. Of the 49 cases, 14 were benign solid lesions (30 images) and 35 were malignant solid lesions (56 images). Finally, the proposed approach will be useful for the analysis of images of the breast of American women and computer-aided diagnostic systems.

15. Architecture of the proposed system

Figure 1 shows the architecture of our system.

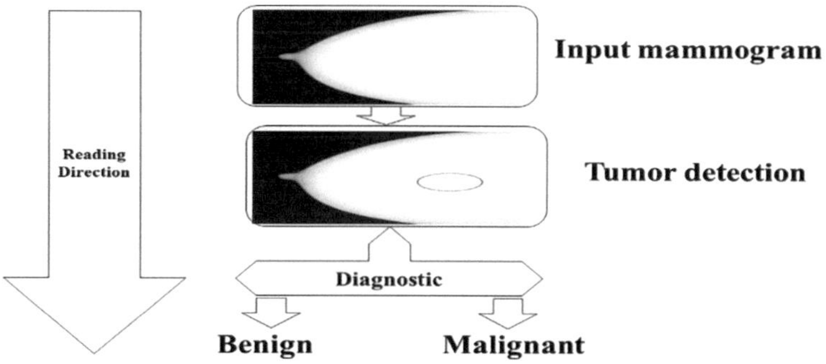

Figure 1.
Architecture of the proposed system (SI2AD for artificial intelligence system aided detection and Diagnosiss).

16. Description of the database used in this study

To carry out our comparative study between artificial intelligence techniques, we chose a breast cancer database extracted from the UCI repository https://archive.ics.uci.edu/ml/datasets/Breast+Cancer+Wisconsin+%28Diagnosi29. This database describes the characteristics of the cell nuclei present in the image. A few of the images can be found at http://www.cs.wisc.edu/~street/images/. Here is the information about this breast cancer database:

- Number of instances: 569
- Number of attributes: 32 (ID, diagnosis, 30 real-valued input features)
- Attribute information:
  - ID number
  - Diagnosis (M = malignant, B = benign) 3–32

Ten real-valued features are computed for each cell nucleus:

a. radius (mean of distances from center to points on the perimeter).

b. texture (standard deviation of gray-scale values).

c. perimeter.

d. area.

e. smoothness (local variation in radius lengths).

f. compactness (perimeter^2/area - 1.0).

g. concavity (severity of concave portions of the contour).

h. concave points (number of concave portions of the contour).

i. symmetry.

j. fractal dimension ("coastline approximation" - 1)

- Class distribution: 357 benign, 212 malignant

17. Performance evaluation measure

$$\text{Accuracy} = \frac{\textit{correct predictions}}{\textit{all predictions}} \; or \; \frac{\textit{True Positive} + \textit{True Negative}}{\textit{all predictions}} \tag{1}$$

18. Results

In this section, we will present our comparative study. We compared artificial intelligence techniques: Logistic Regression, Gradient Boosting Classifier, Random Forest, XGB Classifier, Support Vector Machine, Decision tree, KNeighbors and ANN. The choice of these techniques is based on their very frequent use in the literature.

Artificial Intelligence Methods	Accuracy
ANN	0.97
Decision Tree Classifier	0.94
XGB	0.95
SVM	0.98
Kneighbors Classifier	0.94
Ramdom Forest Classifier	0.95
Gradient Boosting Classifier	0.96
Logistic Regression	0.97

19. Discussion

This work put more emphasis on one of the most powerful algorithms in artificial intelligence. These are convolutional neural networks (CNN). This technique is part of deep learning algorithms. The choice of this method is based on its power, notably allowing the recognition of images by automatically attributing to each image provided as input, a label corresponding to its class of membership. Then, in **Figure 1**, there is the absence of the feature extraction step, this proves that we chose CNN over other AI methods, not by preference but rather on convincing arguments. We know that Artificial Neural Networks (ANNs) like Multilayer Perceptron only contain a classification part, so in systems that use ANNs and want to extract features, it is necessary before applying ANNs before perform a feature extraction step while the CNN contains the two parts: a convolutional part whose final objective is to extract the characteristics specific to each image and a classification part allowing to classify the image. Also, it is sure and certain and all data scientists will tell you that deep learning methods are data and computationally intensive. However, today we see the exponential growth of data in all fields, in particular that of health, in particular with the exponential increase in the number of mammographic images produced. The use and improvement of this method is therefore possible thanks to this big data and the computing power. Here are the essential steps of our proposed system: a tumor detection phase and a diagnostic phase to classify the tumor (benign or malignant). In addition, the proper functioning of our system will be validated in collaboration with experts, in particular doctors, or by using a local database or others. This step is called the validation phase. Finally, our results obtained show that artificial neural networks are efficient, even if the database used in this study is very small.

20. Conclusion

Ultimately, artificial intelligence can play an important role in the early detection of diseases, especially breast cancer. Currently, we are witnessing an exponential increase in large-scale data (or Big Data) in our hospitals. This is caused by the number of medical images produced, the vast majority by women. Thus, in this study, we have shown that it is possible to create robust artificial intelligence systems from medical imaging databases. These systems use machine learning methods in particular deep learning in image classification. In order to facilitate the detection and early diagnosis of breast cancer, we have proposed an aid system called SI2AD as future work.

Author details

Alio Boubacar Goga
Abdou Moumouni University, Niamey, Niger

*Address all correspondence to: gogaalioboubacar@gmail.com

References

[1] Al-masni M, A, Al-antari MA, Park J-m, Gi G, Kim T-Y, Rivera P, et al. Simultaneous detection and classification of breast masses in digital mammograms via a deep learning YOLO-based CAD system. Computer Methods and Programs in Biomedicine. 2018;**157**:85-94

[2] Akbar S, Akram MU, Sharif M, Tariq A, Khan SA. Decision support system for detection of hypertensive retinopathy using arteriovenous ratio. Artificial Intelligence in Medicine. 2018;**90**:15-24. DOI: 10.1016/j.artmed.2018.06.004

[3] Ragab DA, Sharkas M, Marshall S, Ren J. Breast cancer detection using deep convolutional neural network and support vector machines. Peer J. 2019: 1-23. DOI: 10.7717/peerj.6201

[4] Houssein EH, Emam MM, Ali AA, Suganthan PN. Deep and machine learning techniques for medical imaging-based cancer: A comprehensive review. Expert Systems with Applications. 2020: 1-46. DOI: 10.1016/j.eswa.2020.114161

[5] Ting FF, Tan YJ, Sim KS. Convolutional neural network improvement for breast cancer classification. Expert Systems with Applications. 2018;**120**:103-115. DOI: 10.1016/j.eswa.2018.11.008

[6] Qi X, Zhang L, Chen Y, Pi Y, Chen Y, Lv Q, et al. Automated diagnosis of breast ultrasonography images using deep neural networks. Medical Image Analysis. 2019;**52**:185-198. DOI: 10.1016/j.media.2018.12.006

[7] Ceugnart L, Olivier A, Oudoux A. Cancer du sein: la nouvelle imagerie. Presse Médicale. 2019;**48**:1101-1111. DOI: 10.1016/j.lpm.2019.10.007

[8] Sudharshan PJ, Petitjean C, Spanhol F, Oliveira LE. Multiple instance learning for histopathological breast cancer image classification. Expert Systems with Applications. 2019;**117**:103-111. DOI: 10.1016/j.eswa.2018.09.049

[9] Hossam A, Harb HM, Abd HM, Kader E. Automatic image segmentation method for breast cancer analysis using thermography. Journal of Engineering and Science. 2018;**46**:12-32

[10] Tran WT, Jerzak K, Fang-I L, Klein J, Tabbarah S, Lagree A, et al. Personalized breast cancer treatments using artificial intelligence in radiomics and Pathomics. Journal of Medical Imaging and Radiation Sciences. 2019;**50**(4S2):1-10. DOI: 10.1016/j.jmir.2019.07.010

[11] Christine FM, Rémy G. L' intelligence artificielle au service de l'imagerie et de la santé des femmes. 2019;**29**(4):179-186. Available from: https://www.elsevier.com/open-access/userlicense/1.0/

[12] Brunelle F, Brunelle P. Intelligence artificielle et imagerie médicale: définition, état des lieux et perspectives. 2019;**203**:683-687. Available from: https://www.elsevier.com/open-access/userlicense/1.0/

[13] Séminaire FNMR/SFR. L'intelligence artificielle: rêve ou cauchemar du radiologue ?. Juin 2018:1-204

[14] Lassau N, Estienne T, de Vomecourt P, Azoulay M, Cagnol J, Garcia G. Five simultaneous artificial intelligence data challenges on ultrasound, CT, and MRI. Diagnostic and Interventional Imaging. 2019;**100**: 199-209. DOI: 10.1016/j.diii.2019.02.001

[15] Deslandes M, Chave L, Pommier M, Detraz J, Nord B, Panassie L, et al. État de

l'art en imagerie médicale. IRBM News. 2019;**40**(2):45-61

[16] Kedraa J, Gossec L. Big data et analyses par intelligence artificielle: les données de grande ampleur vont-elles modifier notre pratique ?. 2019. Available from: https://www.elsevier.com/open-access/userlicense/1.0/

[17] Gossec L, Kedra J, Servy H, et al. EULAR points to consider for the use of big data in rheumatic and musculoskeletal diseases. Annals of the Rheumatic Diseases. 2020:1-30. DOI: 10.1136/annrheumdis-2019-215694

[18] Kedra J, Radstake T, Pandit A, et al. Current status of use of big data and artificial intelligence in RMDs: A systematic literature review informing EULAR recommendations. RMD Open 5:e001004. 2019:1-8. DOI: 10.1136/rmdopen-2019-001004

[19] Seymoura K, Benyahia N, Hérent P, Malhaire C. Exploitation des données pour la recherche et l'intelligence artificielle: enjeux médicaux, éthiques, juridiques, techniques. Imagerie de la Femme. 2019:1-10. DOI: 10.1016/j.femme.2019.04.004

[20] Frécon L, Kazar O. Manuel d'intelligence artificielle. Presses polytechniques et universitaires romandes. 2009:1-778

[21] Asher N, Bidoit-tollu N, Bloch I, Bonnefon J-f, Bosc B. L'intelligence artificielle: frontières et applications. CÉPADUÈS-ÉDITIONS. 2014

[22] van der Heijden AA, Abramoff MD, Verbraak F, van Hecke MV, Liem A, Nijpels G. Validation of automated screening for referable diabetic retinopathy with the IDx-DR device in the Hoorn Diabetes Care System. Acta Ophthalmologica. 2018;**96**(1):63-68

[23] Durand M, Shaikha A, Billia M, Lechevallier E. Applications médicales de l'intelligence artificielle: opportunités \& challenges. 2020. Available from: https://creativecommons.org/licenses/by-nc/4.0/

[24] Herbreteau D. Enfin l'IA va permettre de remplacer les radiologues. Presse Med Form. 2020;**1**(3):229-230

[25] Lakhani P, Prater AB, Kent Hutson R, et al. Machine learning in radiology: Applications beyond image interpretation. Journal of the American College of Radiology. 2018;**15**(2):350-359. DOI: 10.1016/j.jacr.2017.09.044

[26] Olivier E. Les usages de l'intelligence artificielle. 2018

[27] François C. Deep Learning with Python. Island: Manning Publications; 2018

[28] Bibault JE, Burgun A, Giraud P. Intelligence artificielle appliquée à la radiothérapie. Cancer Radiothérapie. 2017;**21**(3):239-243. DOI: 10.1016/j.canrad.2016.09.021

[29] Dominik F. Machines à vecteurs de support-une introduction. Camus. 2010;**1**:7-25

[30] Or H-S, Zvi B, Shai L, David L, Raphael G, Yehuda Z. Early non-invasive detection of breast cancer using exhaled breath and urine analysis. Computers in Biology and Medecine. 2018:227-232. DOI: 10.1016/j.compbiomed.2018.04.002

[31] Sertan K, Abdulkader H, Dilber U. Breast cancer image classification using artificial neural networks. Procedia Computer Science. 2017;**120**:126-131. DOI: 10.1016/j.procs.2017.11.219

[32] Boukhobza M, Mimi M. Détection automatique de la présence d'anomalie sur une mammographie par réseau de

neurones artificiels. Courrier Du Savoir. 2012;**13**:103-108

[33] Simon HN, Yaecob GG, Abiot SB, Gebrekirstos H. Breast cancer detection using convolutional neural networks. 2020. Available from: https://arxiv.org/pdf/2003.07911.pdf

[34] Fei G, Teresa W, Jing L, Bin Z, Lingxiang R, Desheng S, et al. SD-CNN: A shallow-deep CNN for improved breast cancer diagnosis. Computerized Medical Imaging and Graphics. 2018;**70**:53-62. DOI: 10.1016/j.compmedimag.2018.09.004

[35] Kuo M, Tao Z, Xinyan C, Bo C, Yu T, Shulian W, et al. Fully automatic and robust segmentation of the clinical target volume for radiotherapy of breast cancer using big data and deep learning. Physica Medica. 2018;**50**:13-19. DOI: 10.1016/j.ejmp.2018.05.006

[36] Vijayarajeswari R, Parthasarathy P, Vivekanandan S, Alavudeen BA. Classification of mammogram for early detection of breast cancer using SVM classifier and Hough transfor. Measurement. 2019;**146**:800-805. DOI: 10.1016/j.measurement.2019.05.083

[37] Hiba A, Hajar M, Hassan AM, Thomas N. Using machine learning algorithms for breast Cancer risk prediction and diagnosis. Procedia Computer Science. 2016;**83**;1064-1069. DOI: 10.1016/j.procs.2016.04.224

[38] Ali K, Ayturk K, Ugur Y. Expert system based on neuro-fuzzy rules for diagnosis breast cancer. Expert Systems with Applications. 2011;**38**(5):5719-5726. DOI: 10.1016/j.eswa.2010.10.061

[39] Yanhui G, Cheng HD, Jianhua H, Jiawei T, Wei Z, Litao S, et al. Breast ultrasound image enhancement using fuzzy logic. Ultrasound in Medicine and Biology. 2006;**32**(2):237-247

Chapter 2

Pre-Informing Methods for ANNs

Mustafa Turker

Abstract

In the recent past, when computers just entered our lives, we could not even imagine what today would be like. If we look at the future with the same perspective today, only one assumption can be made about where technology will go in the near future; Artificial intelligence applications will be an indispensable part of our lives. While today's work is promising, there is still a long way to go. The structures that researchers define as artificial intelligence today are actually programmed programs with limits and are result-oriented. Real learning includes many complex features such as convergence, association, inference and prediction. It has been demonstrated with an application how to transfer the input layer connections in human neurons to the artificial learning network with the pre-informing method. When the results are compared, the learning load (weights) was reduced from 147 to 9 with the proposed pre-informing method, and the learning rate was increased between 15–30% according to the activation function used.

Keywords: ANN, pre-informing, AHP, modified networks, interfered networks

1. Introduction

The learning mechanism makes human beings superior to all other creatures. Despite the fact that today's computers have much more processing power, the human brain is still much more efficient than any computer or any artificially developed intelligence.

Building a perfect learning network requires more than just cell structures and its weights. The human brain has a very complex network, and each brain is unique for itself. Today's technology is not enough to explain all the details of how our brain works. My observation of how our brain works starts from defining items. Every item has a key cell in our brain. Defining process is done by visuals, smell, feeling, linguistic name, hearing its sound. If these key cells match any of these information from body inputs, thinking and learning continues, if there is no key cell defined before, new cell is assigned for this item. Then, your brain wants to explore these item's behavior. You start to take this item in your hand and start the psychical observation. When the psychical observation is satisfied, your brain starts to categorize it. After categorization, your brain checks other items for same categorization and determines what other information can be learned. Whenever you see someone has more knowledge from you, then you want to speak about this newly learned item, or you want to do research on it. This key cell started to develop itself with explored information. Each key cell

IntechOpen

and its network can also connect to each other in any part, if there are logical connections that exist.

Today's artificial intelligence studies are a little simple compared to reality. Mathematical modeling of learning in an artificial cell and solving the problem with an optimization mechanism has resulted in success in most areas. However, this success is due to the fast processing capacity of computers rather than the perfect modeling of machine learning. In this case, researchers need to work on developing artificial neural networks close to the real learning.

In this study, the pre-informing method and rules in artificial neural networks are explained with an example in order to establish a more conscious and effective learning network instead of searching for relationships in random connections.

2. ANN structure

In the literature of ANN design, the first principles were introduced in the middle of the 20th century [1, 2]. Over the following years, network structures such as Perceptron, Artron, Adaline, Madaline, Back-Propagation, Hopfield Network, Counter-Propagation Network, Lamstar were developed [3–10].

The complex behavior of our brain artificially imitated through layers is most network configuration. Basically, an artificial neural network has 3 types of layer group: input layer, hidden layers, output layer (See **Figure 1**). And all cells in these layers connected each other with artificial weights [1, 2].

Input layer is the cluster of cells present the data that has influence on learning. Each cell represents a parameter with a variable data value. These values are scaled

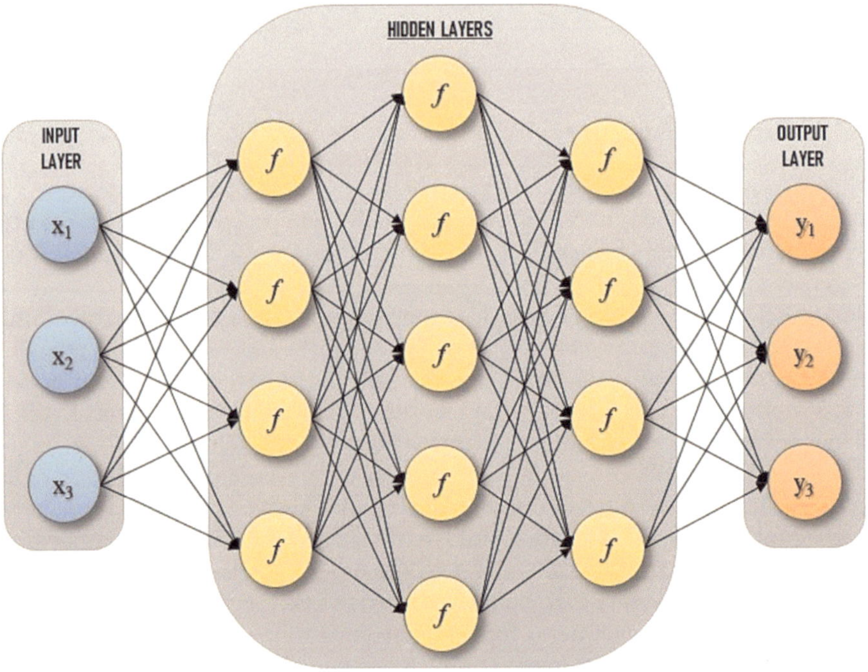

Figure 1.
Basic ANN structure.

according to the limits of the activation function used in the next layers. The selection of input parameters requires knowledge and experience on the subject to be created artificial intelligence. In fact, this process is exactly the transfer of natural neuron input parameters from our brain to paper. However, this is not so easy because a learning activity in our brain is connected by a huge number of networks managed subconsciously. To explain this situation, sometimes our minds make some inferences even on subjects we have no knowledge of, and we can make correct predictions about this subject. In some cases, we feel the result of an event that we do not know, but we cannot explain it. In fact, the best example of this is falling in love. No one can tell why you fall in love with a person, it happens and then you look for the reason. This is proof that the subconscious mind plays a major role in learning. This means that there may also be some input parameters that we did not notice. Therefore, it is necessary to focus on this layer and define the input parameters.

Hidden layer(s) is the layer where the data of the input parameters are interpreted, and the learning capability of the network is defined. Each cell in these layers transfers the data from the input layer cells or previously hidden layer cells with the defined activation function and sends it to all cells in the next layer. Learning of nonlinear behavior takes place in this layer. Increasing the number of layers and cells in this group does not always work, but provides memorization, not learning. This also increases the number of connections and thus highly increases the required experienced data to determine the weight values of these connections.

In general, the basic mechanism of an artificial neuron consists of two steps: summation and activation [1]. Summation is the process of summing the intensities of incoming connections. Activation, on the other hand, is the process of transforming the collected signals according to the defined function (See **Figure 2**).

There are many activation functions. The purpose of these functions is to emulate linear or non-linear behavior. The sigmoid function is one of the most commonly used activation functions.

Mathematically, the summation and activation process of an artificial neuron is expressed as below (See Eqs. (1) and (2)).

$$u = \sum_{n=1}^{m} x_i * w_i - \theta \tag{1}$$

$$y = f(u) \tag{2}$$

In these equations,

- x_i: Input value or previous cell output value for previous layer cells,
- w_i: Weight value of the connection for previous layer cells,

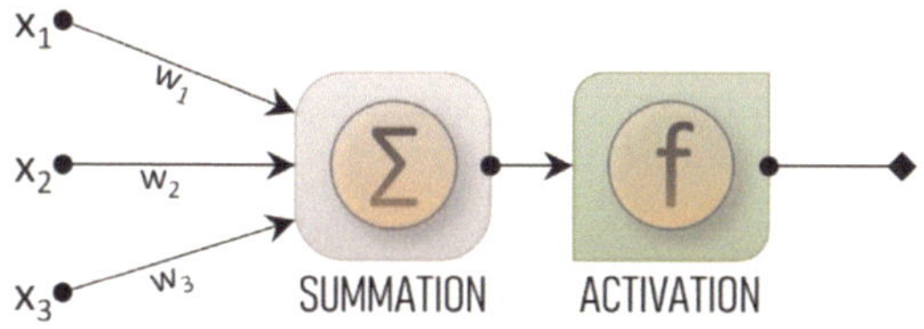

Figure 2.
Artificial neuron structure.

- θ: Bias value,
- u: Net collected output value of the cell,
- y: Activated output value of the cell.

In some cases, the learning network cannot find a logical connection between the results and the inputs, so that this does not stop learning, a bias value can be used for each cell. A high bias coefficient means that learning is low, and memorization is high.

The output layer is the last layer in the connection and receives inputs from the last set in the hidden layer. In this layer, data is collected and as a result, output data is exported in the planned method.

The learning process of the network established with the input, hidden and output layers is actually an optimization problem. The connection values between the cells of the network converge to reach the result depending on the optimization technique. A training set consisting of a certain number of input and output data is used for this purpose. If desired, a certain amount of data set is also tested to measure the consistency of the network. When the learning is complete, the values of the weights are fixed, and the network becomes serviceable. If desired, the mathematical equation of the network can be derived by following the cells from back to forward.

3. Pre-informing of ANNs

Pre-information, unlike pre-training, is the processing of a certain information or rule into the structure of the network. In reality, a person learns under some prejudices while learning something. These prejudices are a mechanism that allows us to make predictions about the event that will occur, and they make these inferences by utilizing similar events. With these prejudices, the number of training data required for learning decreases by a considerable ratio. As a result, you have a clean and efficient way of learning.

For example, for a child who goes out for the first time, his mother advises never to talk to strangers, and he guesses that if the child talks to a stranger, the result may be bad. In this case, the people to talk to are the input parameters, the possibility of something bad happening as a result of the conversation is the output parameter. If the mother did not give advice to her child, the child would talk to everyone and eventually learn that talking to a stranger is bad and dangerous. As a result of the mother's suggestion, the weight of strangers among the input parameters (people to talk to) increased before they even experienced it.

In order to transfer prejudices to artificial neural networks, some rules must be followed:

1. The pre-Informed network structure consists of 3 layers; input layer, hidden layer, output layer. The hidden layer consists of a single sublayer.

2. Input parameters should be grouped, if possible. For example, in a learning network that predicts heart attack, personal characteristics are one group, bad habits are another group, genetic diseases are another group. If there is no group, it should be considered as 1 group. These inputs should be scaled according to the activation function that will be used in the hidden layer.

3. The information to be processed (pre-informing) should be in the weights between the input layer and the hidden layer.

4. An artificial neuron cell is placed for each input group in the hidden layer to represent each group. This cell consists of 3 steps: summation, scaling, activation. Two or more different activation functions can be used in cells in the hidden layer. In this case, for each input group, same number of representation cells should be defined in the hidden layer.

5. The connections of cells in the input layer to the representation cells of other groups other than their own are considered 0.

6. The representation cells in the hidden layer are directly connected to the output layer.

7. Optimization optimizes the weights of the connections between the hidden layer cells and the output layer cells.

8. The connection values of the input layer groups to the representation cells in the hidden layer are determined and fixed for each group using the techniques in the literature.

In **Figure 3**, a total of 23 input parameters belonging to 3 input groups, these three groups are represented by two separate cells with hyperbolic tangent and sigmoid activations, and a hidden layer consisting of a total of 6 cells, and finally an output layer are described.

After the network structure is established, the next step is pre-informing the network. This stage is the transfer of information from the subconscious to network weights. This stage should be done for each group, and each group should be considered separately. The best method of this process is using AHP (Analytic Hierarchy Process) evaluation methods. In AHP evaluation methods, each parameter is compared with the other using verbal expressions. A simple superiority scale is used in this comparison. This means you can prepare a questionnaire and get the superiority information of parameters from an expert mind. After some calculations you will have the weights. These weights will be used in the network directly. The beauty of using this technique is consistency analysis can be done. In the end, if the input parameters are defined correctly, you will have 100% academically proofed subconscious information extraction.

AHP is a multi-criteria decision making (MCDM) method. The earliest reference to AHP is from 1972 [11]. Afterwards, Saaty [12], fully described the method in his article published in the Journal of Mathematical Psychology. AHP makes it possible to divide the problem into a hierarchy of sub-problems that can be more easily grasped and evaluated subjectively. Subjective evaluations are converted into numerical values and each alternative is processed and ranked on a numerical scale. Schematic AHP hierarchy is given in **Figure 4** below.

At the top of the hierarchy is the goal/purpose, while at the bottom there are alternatives. Between these two parts are the criteria and their sub-criteria. The most important feature that makes AHP important is that it can make comparisons both locally and globally when comparing the effect of sub-criteria at any level on alternatives.

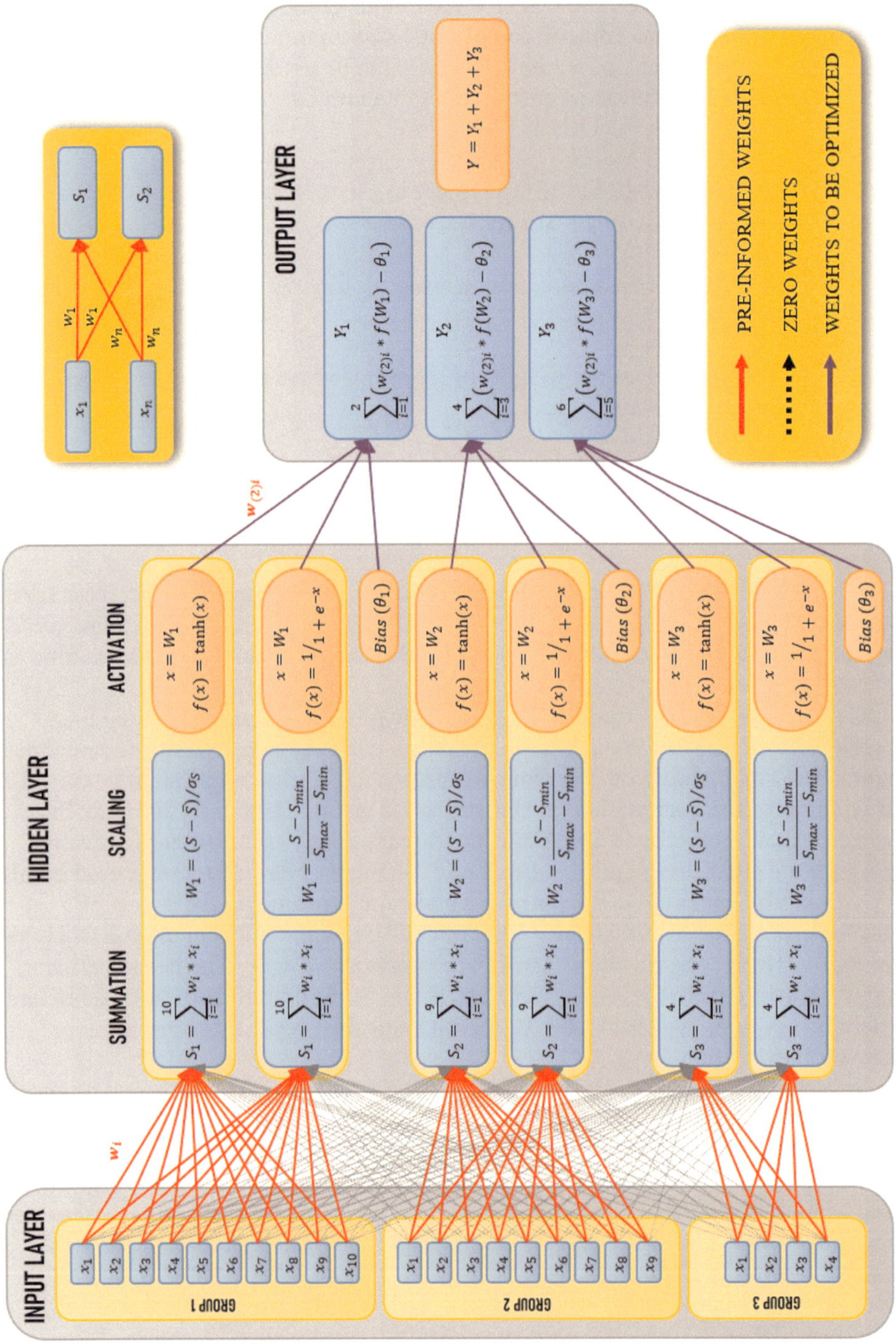

Figure 3.
Pre-informed ANN structure.

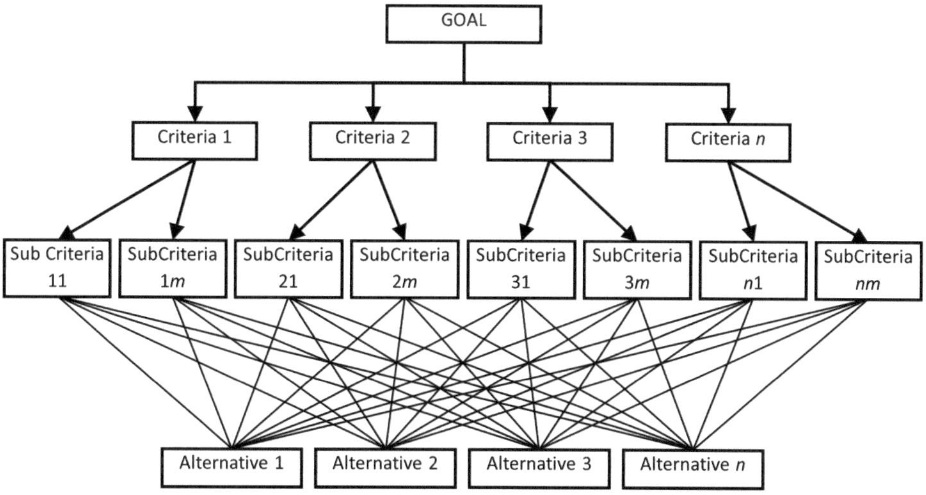

Figure 4.
AHP hierarchy.

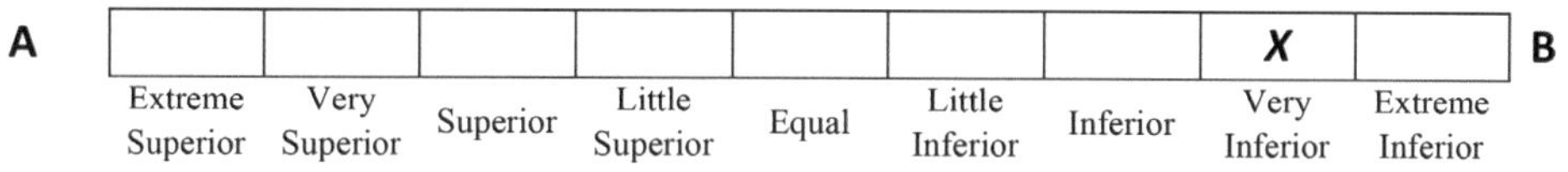

Figure 5.
Pairwise comparison chart of alternatives A and B. B is very inferior compared to A.

Data corresponding to the hierarchical structure is collected by experts or decision makers by pairwise comparison of alternatives within the scope of a qualitative scale. Experts can rate the comparison as equal, less strong, strong, very strong and extremely strong. A general table, as shown in **Figure 5**, is used for expert evaluation of pairwise comparisons and data collection. This design can be customized for purpose, method and user usage.

Comparisons are made for each criterion and converted to quantitative numbers according to **Table 1**.

The pairwise comparison values of the criteria arranged in a matrix shown in **Table 2**.

Scale	Definition	Description
1	Equal	The two criteria are equally important.
3	Little Superior	One of the criteria has some superiority based on experience and judgment
5	Superior	One of the criteria has many advantages based on experience and judgment.
7	Very Superior	One criterion is considered superior to the other
9	Extreme Superior	Evidence that one criterion is superior to another has great credibility
2, 4, 6, 8	Intermediate values	Intermediate values to be used for reconciliation

Table 1.
Comparison scales and explanations.

	C_1	C_2	C_3	C_n
C_1	$a_{11} = 1$	a_{12}	a_{13}	a_{1n}
C_2	$1/a_{12}$	$a_{22} = 1$	a_{23}	a_{2n}
C_3	$1/a_{13}$	$1/a_{23}$	$a_{33} = 1$	a_{3n}
C_n	$1/a_{1n}$	$1/a_{2n}$	$1/a_{3n}$	$a_{nn} = 1$
$\sum a$	$S_1 = \sum_{i=1}^{n} a_{i1}$	$S_2 = \sum_{i=1}^{n} a_{i2}$	$S_3 = \sum_{i=1}^{n} a_{i3}$	$S_n = \sum_{i=1}^{n} a$

Table 2.
Pairwise comparison matrix of criteria.

	K_1	K_2	K_3	K_n	w_i
K_1	a_{11}/S_1	a_{12}/S_2	a_{13}/S_3	a_{1n}/S_n	w_1
K_2	a_{21}/S_1	a_{22}/S_2	a_{23}/S_3	a_{2n}/S_n	w_2
K_3	a_{31}/S_1	a_{32}/S_2	a_{33}/S_3	a_{3n}/S_n	w_3
K_n	a_{n1}/S_1	a_{n2}/S_2	a_{n3}/S_3	a_{nn}/S_n	w_n
$\sum a$	S_1/S_1	S_2/S_2	S_3/S_3	S_n/S_n	$w_i = \left[\sum_{j=1}^{n}\left(\frac{a_{ij}}{S_j}\right)\right]/n$

Table 3.
Obtaining the weights of the normalized comparison values of the criteria.

In next step, each a_{ij} value is normalized by dividing by the corresponding column sum, and the weights shown in the table above are obtained with the corresponding equation shown in the **Table 3** above.

Network connections of input parameters using AHP are explained as shown above. Next step is how to assign weights. **Figure 6** shows how the AHP weights are defined to the network.

In this way, a large number of connections are canceled and a fast, efficient and less data-needing network is obtained.

4. Estimation of the severity of occupational accidents with using pre-informed ANN

The pre-informed neural network method was used by Turker [13] to predict the severity of occupational accidents in construction projects. In this study, it has been estimated how the accidents will result if they happen instead of the possibility of their occurrence. The scope of the study was made for the 4 most common accident types in the world. These are falling from high, hit from a thrown/falling object, structural collapse, electrical contact. In this study, 23 measures to be taken in occupational accidents are discussed in 3 groups. These measures have been associated with occupational accident severity in the artificial intelligence network (**Table 4**).

First of all, defined measures in occupational accidents, which are the input parameters, were turned into a questionnaire by creating paired comparison questions for comparison within their own groups. Occupational health and safety experts working professionally in the sector were reached through a professional firm. The questionnaires were administered online and recorded. Survey results were taken and converted to weights with AHP matrices. Weights are shown in **Tables 5–7**.

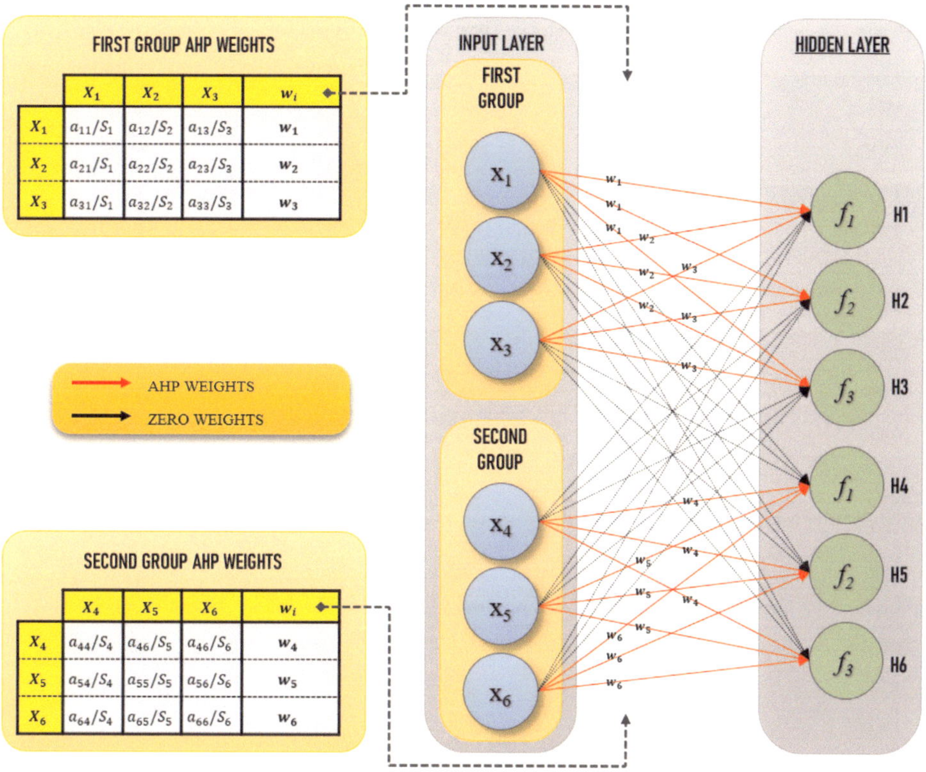

Figure 6.
Connections of two input groups to three different types of representation cells and implementation of AHP weights.

Collective protection measures (TKY)	Personal protective equipment (KKD)	Control, training, inspection (KEM)
(TKY-1) Constr. site curtain system	(KKD-1) Safety Helmet	(KEM-1) OHS specialist
(TKY-2) Colored excavation net	(KKD-2) Protective Goggles	(KEM-2) Occupational Doctor
(TKY-3) Safety rope system	(KKD-3) Face Mask	(KEM-3) Examination
(TKY-4) Guardrail systems	(KKD-4) Face Shield	(KEM-4) OHS trainings
(TKY-5) Facade cladding	(KKD-5) Working Suit	
(TKY-6) Safety Field Curtain	(KKD-6) Reflector	
(TKY-7) First aid kit, fire extinguisher	(KKD-7) Parachute Safety Belt	
(TKY-8) Facade safety net	(KKD-8) Working Shoes	
(TKY-9) Mobile electrical dist. panel	(KKD-9) Protective Gloves	
(TKY-10) Warning and info signs		

Table 4.
Risk reduction measures in occupational accidents.

After obtaining the preliminary information weights, 3 different artificial intelligence networks were created (**Table 8**). 140 historical accident data were collected on selected accidents within a company. These data include the precautions taken at the time of the accident and how the accident resulted. Accident results are divided into 4 categories: near miss, minor injury, serious injury, death. For each accident,

Code	Structural collapse	Falling from high	Object hit	Contact w/ Electricity
TKY-1	0,000	0,000	0,000	0,000
TKY-2	0,000	0,000	0,000	0,000
TKY-3	0,555	0,398	0,109	0,000
TKY-4	0,000	0,185	0,109	0,000
TKY-5	0,000	0,102	0,000	0,000
TKY-6	0,252	0,099	0,109	0,107
TKY-7	0,097	0,039	0,406	0,120
TKY-8	0,000	0,126	0,000	0,000
TKY-9	0,000	0,000	0,000	0,411
TKY-10	0,097	0,052	0,269	0,361

Table 5.
AHP weights of collective protection measures group.

Code	Structural collapse	Falling from high	Object hit	Contact w/ electricity
KKD-1	0,195	0,243	0,225	0,076
KKD-2	0,095	0,050	0,080	0,098
KKD-3	0,044	0,044	0,035	0,039
KKD-4	0,072	0,050	0,091	0,100
KKD-5	0,071	0,093	0,106	0,179
KKD-6	0,039	0,044	0,050	0,058
KKD-7	0,337	0,388	0,252	0,086
KKD-8	0,081	0,045	0,087	0,142
KKD-9	0,066	0,045	0,074	0,222

Table 6.
AHP weights of personal protective equipment group.

Code	Structural collapse	Falling from high	Object hit	Contact w/ electricity
KEM-1	0,481	0,167	0,399	0,426
KEM-2	0,210	0,167	0,161	0,134
KEM-3	0,098	0,167	0,083	0,067
KEM-4	0,210	0,500	0,357	0,372

Table 7.
AHP weights of control, training, inspection group.

35 datasets were collected and a total of 120 datasets were used in training the network and 20 datasets were used in testing the network.

Three alternative network structures were trained with the same data. As a result, the pre-informed neural network provided a better learning rate of 5% in the training

Network	Regular ANN	Pre-informed ANN	Pre-informed ANN
Software	SPSS – Neural Networks Engine	EXCEL VBA + SOLVER	EXCEL VBA + SOLVER
Network Structure	Multilayer Perceptron (MP)	MP	MP
Number of Hidden Layers	1	1	1
Cells in Hidden Layer	6 (Cells) + 1 (Bias)	6 (Cells) + 3 (Bias)	6 (Cells) + 3 (Bias)
Activation Function in Hidden Layer Cells	Hyperbolic Tangent (6 Cells)	Hyperbolic Tangent (6 Cells)	Parabolic Functions 3 Cells; $f(x) = x^2$ 3 Cells; $f(x) = x$
Output function	$f(x) = x$	$f(x) = x$	$f(x) = x$
Scaling Method	(x- x̄) / Standard Dev.	(x- x̄) / Standard Dev.	(1–x) * 10
Optimization Algorithm	Gradient Methods	Gradient Methods	Gradient Methods
Randomizer	Mersenne Twister algorithm	Mersenne Twister	Mersenne Twister
Initial Value	10	10	0,1

Table 8.
3 alternative ANN structures.

set and 15% in the test set compared to the neural network without a pre-informed stage. The configuration using parabolic activation function from pre-informed artificial neural networks provided 1% better learning rate in the training set and 15% better in the test set compared to the configuration using hyperbolic tangent. Other configurations with activation functions were not included in the comparisons because of their low learning rates. As a result, it has been seen that the preliminary information phase significantly increases the learning performance in artificial neural networks. In addition, it has been observed that the parabolic activation function performs better than the hyperbolic tangent in relation to the prevention methods in occupational accidents and the result of the accident (**Table 9**).

	Network	Regular ANN	Pre-informed ANN	Pre-informed ANN
STRUCTURAL COLLAPSE	Training Set	26/30 (87%)	29/30 (97%)	30/30 (100%)
	Test Set	2/5 (40%)	4/5 (80%)	4/5 (80%)
CONTACT w/ ELECTRICITY	Training Set	27/30 (90%)	30/30 (100%)	30/30 (100%)
	Test Set	2/5 (40%)	4/5 (80%)	5/5 (100%)
OBJECT HIT	Training Set	30/30 (100%)	30/30 (100%)	30/30 (100%)
	Test Set	4/5 (80%)	3/5 (60%)	5/5 (100%)
FALLING FROM HIGH	Training Set	30/30 (100%)	30/30 (100%)	30/30 (100%)
	Test Set	4/5 (80%)	4/5 (80%)	4/5 (80%)
TOTAL	Training Set	113/120 (94%)	119/120 (99%)	120/120 (100%)
	Test Set	12/20 (60%)	15/20 (75%)	18/20 (90%)

Table 9.
3 alternative ANN structure results.

5. Conclusions

In this study, how the learning ability of artificial neural networks should be increased with the pre-informing method is explained with rules and demonstrations. It is not possible to implement this method with the existing ready-made ANN software on the market. Instead, ANN should be expressed mathematically, and pre-informing method should be applied using programming languages such as MATLAB, Excel VBA, Python.

In this section, the application of this method has been demonstrated in an artificial neural network in which the precautions in occupational accidents are associated with the results of the accident and high performance has been achieved. With the application of the specified rules, this method can be used to solve many problems. In future studies, it can be investigated which other methods such as AHP can be used for the preliminary information phase.

Conflict of interest

The authors declare no conflict of interest.

Author details

Mustafa Turker
Gorkem Construction Company, Ankara, Turkiye

*Address all correspondence to: mustafaturker@yahoo.com

References

[1] Graupe D. Principles of artificial neural networks. 3rd ed. In: Advanced Series in Circuits and Systems. Singapore: World Scientific Publishing Co. Pte. Ltd.; 2013. DOI: 10.1142/8868

[2] McCulloch WS, Pitts W. A logical calculus of the ideas immanent in nervous activity. The Bulletin of Mathematical Biophysics. 1943;**5**(4): 115-133

[3] Rosenblatt F. The perceptron: A probabilistic model for information storage and organization in the brain. Psychological Review. 1958; **65**(6):386

[4] Graupe D, Lynn J. Some aspects regarding mechanistic modelling of recognition and memory. Cybernetica. 1969;**12**(3):119

[5] Hecht-Nielsen R. Counterpropagation networks. Applied Optics. 1987;**26**(23): 4979-4984

[6] Hopfield JJ. Neural networks and physical systems with emergent collective computational abilities. Proceedings of the National Academy of Sciences. 1982;**79**(8): 2554-2558

[7] Bellman R, Kalaba R. Dynamic programming and statistical communication theory. Proceedings of the National Academy of Sciences. 1957; **43**(8):749-751

[8] Widrow B, Winter R. Neural nets for adaptive filtering and adaptive pattern recognition. Computer. 1988; **21**(3):25-39

[9] Widrow B, Hoff ME. Adaptive Switching Circuits. Stanford, CA: Stanford University; 1960

[10] Lee RJ. Generalization of learning in a machine. In: Preprints of Papers Presented at the 14th National Meeting of the Association for Computing Machinery (ACM '59). New York, NY, USA: Association for Computing Machinery; 1959. pp. 1-4. DOI: 10.1145/612201.612227

[11] Saaty TL. An Eigenvalue Allocation Model for Prioritization and Planning. Pennsylvania, USA: University of Pennsylvania; 1972. pp. 28-31

[12] Saaty TL. A scaling method for priorities in hierarchical structures. Journal of Mathematical Psychology. 1977;**15**(3):234-281

[13] Turker M. Estimation of the Severity of Occupational Accidents in the Building Process with Pre-informed Artificial Learning Method. Gazi: Gazi University; 2021

Chapter 3

Artificial Neural Network Logic-Based Reverse Analysis with Application to COVID-19 Surveillance Dataset

Hamza Abubakar and Muntari Idris

Abstract

The Boolean Satisfiability Problem (BSAT) is one of the crucial decision problems in the fields of computing science, operation research, and mathematical logic that is resolved by deciding whether or not a solution to a Boolean formula exists. When there is a Boolean variable allocation that induces the Boolean formula to yield TRUE, then the SAT instance is satisfiable. The main purpose of this chapter is to utilize the optimization capacity of the Lyapunov energy function of Hopfield neural network (HNN) for optimal representation of the Random Satistibaility for COVID-19 Surveillance Data Set (CSDS) classification with the aim of extracting the relationship of dominant attributes that contribute to COVID-19 detections based on the COVID-19 Surveillance Data Set (CSDS). The logical mining task was carried based on the data mining technique of the energy minimization technique of HNN. The computational simulations have been carried using the different number of clauses in validating the efficiency of the proposed model in the training of COVID-19 Surveillance Data Set (CSDS) for classification. The findings reveals the effectiveness and robustness of k satisfiability reverse analysis with Hopfield neural network in extracting the dominant attributes toward COVID-19 Surveillance Data Set (CSDS) logic.

Keywords: artificial neural network, Hopfield neural network, random satisfiability, reverse analysis, logic mining

1. Introduction

The COVID-19 pandemic is still having a significant impact on people's health and quality of life all around the world. Effectively identifying and isolating affected people is the most crucial step in stopping COVID-19. Clinical medicine can identify COVID-19 instances thanks to the discovery that patients with the infection exhibit anomalies in CT imaging. Additionally, CT scans can be used to determine the severity of an illness, which is useful for selecting the right course of action. Many deep learning-based COVID-19 case identification techniques have recently been put out, some of which have had good success. The biggest obstacle to increasing COVID-19's classification accuracy is currently the small number of training instances and annotations. Additionally, due to the poor contrast of CT scans, deep learning-based classification systems struggle to comprehend ambiguous and imprecise information, such as pixels near boundaries and pictures related to COVID-19 instances. We suggest a belief function-based classification network to categorize COVID-19 cases using semi-supervised learning to address the aforementioned issues, and we acknowledge the research community's open-source COVID-19 dataset [1–3].

Neural networks are used extensively in many fields of study originated in mathematical neurobiology. This can be because these networks are attempts to simulate human brain capabilities. Neural networks have been utilized since the last decade as a theoretically sound alternative to traditional statistical models. It is also effective in classifying data into identifiable groups or characteristics. When used in a hybrid framework with the many forms of predictive neural networks, the classification of neural networks becomes very efficient. Machine-learning methods including artificial neural networks (ANN) have been used in recent times as tools for decision, prediction, classification, and diagnosis [4–11]. It has been used in medical approaches to digenesis, predict, and detect diseases using effects on development, such as fibrosis prediction, cirrhosis, and prediction of response therapy in patients with hepatitis C [12–21].

Artificial neural network (ANN) models have been widely used in data mining for a wide range of medical, science, engineering, and industry issues. Logic mining is one of the key data mining fields. It has been shown that knowledge in a logical or symbolic form can be represented [22]. Recent data mining studies have led to the emergence of various types of artificial neural network models such as Hopfield neural network (HNN), Radial Basis Neural Network, Convolution neural network, and other machine learning tools that can foster logic mining through knowledge extraction process [23, 24]. Consequently, data mining practitioners assimilate multidisciplinary knowledge such as Artificial neural networks, mathematics, artificial intelligence, machine learning, and statistics to create logic for data mining techniques for finding underlying information based on the behavior of databases or data sets. Therefore, data mining can be improved in the neural network to cater to various problems. As for this work, the incorporation of Random *k* Satisfiability (RANkSAT) propositional logic is utilized with the proposed to introduce a comprehensive model to solve real-life applications. Artificial neural network (ANN) possesses a comprehensive structure of training and testing stages thus, ANN has emerged as one of the most efficient tools extend in finding patterns and extracting information to solve real-life applications. Therefore, we presume that this research contributes to amplifying the efforts to enhance the capability of fundamental ANN with the inclusion of a recurrent type of network known as Hopfield neural network (HNN). Propositional logic based on Random-k Satisfiability (RANkSAT) is considered a suitable approach to represent logical rules in neural network for optimal classification of real-world problem. By considering only maximum of three literals per clause, the logical complexity in learning the relationship between the variables in real-life problem decreases.

In this work, Random *k* Satisfiability-based Reverse Analysis method (RAN*k*SATRA) has been proposed to obtain the logical COVID-19 Surveillance Data Set (CSDS). The aforementioned studies, which offer a different perspective on describing the real data in the form of logical representation, have demonstrated the usefulness of diverse logic programming in HNN. In order to depict the behavior of the COVI-19 data sets, we require a very well, intelligent logical rule. However, there is no effort to bridge RANkSAT logical representation in Hopfield neural network for COVID-19 classification. This is crucial because an artificial neural network algorithm can cater variation and randomness in COVID-19 analysis and larger searching space. Therefore, the contributions of this research are presented as follows: The contributions of this work are as follows: (a) to convert the COVID-19 surveillance data set (CSDS) into intelligent systematic form based on RAN*k*SAT logical clauses. (b) To propose random *k* satisfiability reverse analysis method as an alternative approach in extracting the relationships between the factors or attributes that contribute to the knowledge extraction based on COVID-19 surveillance data set (CSDS) obtained from a UCI machine learning repository. (c) To assess the capability and accuracy of three variants of the satisfiability proposed method based on Random k satisfiability, Random maximum k satisfiability, and Horn k satisfiability logical representation in completing the COVID-19 surveillance data set (CSDS) extraction with a different number of clauses. The performance evaluation metrics will be adopted to evaluate the effectiveness of both the proposed method and logical representations as an alternative data extraction method to the COVID-19 data set. The general implementation of random *k* satisfiability reverses the analysis method and HNN in extracting logic in COVID-19 data. The construction of our proposed model, would exhibit better performance in the training stage and successfully interprets real-life datasets to detect which factors are more prominent than others that contribute to the optimization problems. Our findings showed that the proposed model executing the best performance in terms of attaining small errors and efficient computational time compared to other existing models. This study has been organized as follows. Material and method including, Random *k* Satisfiability Logic, Hopfield Neural Network (HNN), and random *k* satisfiability-based reverse analysis method (RAN*k*SATRA) have been described in Section 2. In Section 3 Implementation procedure for classifying the COVID-19 data set was presented. Section 4 reported model simulations and experimental setup, and section. 5 reported performance evaluation metrics Section 6 presented the result and discussion. The chapter concludes with future works presented in Final Section.

2. Materials and methods

2.1 Random k satisfiability logic

Propositional satisfiability logic can perceive as a logical rule that consists of clauses that contain literals or variables. Random *k* Satisfiability (RAN*k*SAT) is a class of non-systematic Boolean logic representation, consists of a random number of literals (can be the negated literals) per clause. Non-systematic Boolean Satisfiability logic (RAN*k*SAT) has been proven effective to represent simulated applications [25]. There is no study that utilizes the non-systematic behavior of Random *k* Satisfiability in discrete HNN for application real data set classification problems. The formulation of RAN*k*SAT has the following properties:

a. A set of variables$\{x_1, x_2, x_3, \ldots, x_n\}$, in a clause C_i where $i = 1, 2, 3, \ldots, n$ that consists of $x\ or \neg x$.

b. Randomly select 2 variables from the set of n variables with a 0.5 probability of negating each variable in the clause.

c. Each x_i in C_i is connected by a disjunction($\vee$).

d. Each clause C_i is connected by a conjunction($\wedge$).

The Boolean values for each x_i are bipolar$x_i \in \{-1, 1\}$ that exemplifies the notion of FALSE and TRUE respectively. The general formulation $F_{RAN3SAT}$ is represented in Eq. (1) as follows.

$$F_{RANkSAT} = \wedge_{i=0}^{t} C_i^{(3)} \wedge_{i=0}^{n} C_i^{(2)} \wedge_{i=0}^{m} C_i^{(1)} \tag{1}$$

where $t, n, m \in [1, 2, ..k], \forall t, n, m > 0$. The clause $F_{RAN3SAT}$ is defined as a random 3-SAT which consists of a clause $C_i^{(k)}$ described in Eq. (2). as follows.

$$C_i^{(k)} = \begin{cases} (\tau_i \vee \alpha_i \vee \Psi_i), k = 3 \\ (\tau_i \vee \alpha_i), \quad k = 2 \\ \lambda_i, \quad k = 1 \end{cases} \tag{2}$$

where $\tau_i \in [\tau_i, \neg\tau_i]$, $\alpha_i \in [\alpha_i, \neg\alpha_i]$, $\Psi_i \in [\Psi_i, \neg\Psi_i]_,$ and $\lambda_i \in [\lambda_i, \neg\lambda_i]$ represent literals and their negation respectively. In particular, the first and second-order clauses are denoted as $C_i^{(1)}$, $C_i^{(2)}{}_,$ and $C_i^{(3)}{}_,$ respectively. In this chapter, F_r is a Conjunctive Normal Form (CNF) formula where the clauses are chosen uniformly, independently without replacement among all $2^r \begin{pmatrix} m+n \\ v \end{pmatrix}$ nontrivial clauses of the length r. Note that, A_i exists in the $C_i^{(k)}$, if the $C_i^{(k)}$ contains either G_i or $\neg G_i$ and the mapping of $V(F_r) \rightarrow \{-1, 1\}$ is called logical interpretation. The Boolean value for the mapping is expressed as 1 (TRUE) and $-$ 1 (FALSE). In theory, the example of RAN*k*SAT formula for $k \leq 3$ is given as.

$$F_{RAN3SAT} = (\tau_1 \vee \neg\tau_2 \vee \tau_3) \wedge (\neg\alpha_1 \vee \alpha_2) \wedge \neg\lambda_1 \tag{3}$$

According to Eq. (3), $F_{RANkSAT}$ comprises of Eq. (4)–(6) as follows.

$$C_i^{(3)} = (\tau_1 \vee \neg\tau_2 \vee \tau_3) \tag{4}$$

$$C_2^{(2)} = (\neg\alpha_1 \vee \alpha_2) \tag{5}$$

$$C_1^{(1)} = \neg\lambda_1 \tag{6}$$

Therefore, the outcome of Eq. (3) is satisfied if.
Eqs. (4)–(6) are satisfied. i.e.,

$$C_i^{(3)} = C_1^{(2)} = C_2^{(1)} = 1 \tag{7}$$

In this study, $F_{RANkSAT}$ will be embedded in the proposed model based on reverse analysis technique for COVID-19 data classification. $F_{RANkSAT}$ will cater the modified networks to unveil the true pattern or behavior of the real data sets involved. Note that $F_{RANkSAT}$ is a symbolic form representation thus it is appropriate to be integrated in these networks as HNN is a nonsymbolic platform.

2.2 Hopfield neural network (HNN)

Hopfield type of neural network (HNN) is a recurrent neural network (RNN) that mimics the human biological brain learning structure. The architecture of HNN model consists of interconnected neurons and a powerful feature of content addressable memory that is crucial in solving various optimization and combinatorial tasks [26]. The system consists of structured N neurons, each of which is represented by an Ising variable. The neurons in discrete HNN are utilized in bipolar representation whereby $S_i \in \{1, -1\}$, which strictly considers values of 1 and -1 [27]. The fundamental overview for neuron state activation in HNN is shown in Eq. (4)

$$S_i = \begin{cases} 1 \ , \ if \ \sum\limits_j T_{ij} S_j > \omega \\ -1 \ , Otherwise \end{cases} \tag{8}$$

where T_{ij} is the synaptic weight from unit j to i. S_j is the state of neuron j and ω is the predefined threshold value. Barra et al. (2018) specified that $\omega = 0$ to certify the network's energy decreases. The connection in Hopfield net contains no connection with itself as follows.

$$T^{(3)}_{ijk} = T^{(3)}_{kij} = T^{(3)}_{kji} \tag{9}$$

$$T^{(2)}_{ji} = T^{(2)}_{ij} \tag{10}$$

$$T^{(1)}_{i} = T^{(1)}_{j} \tag{11}$$

$$T_{jj} = T_{ii} = 0 \tag{12}$$

In resulting, HNN holds symmetrical features in terms of architecture. HNN model has similar intricate details to the *Ising* model of magnetism [28]. As the neuron state is termed in bipolar $S_i \in \{1, -1\}$ representation, the spin points follow in the direction of a magnetic field. This causes each neuron to flip until the equilibrium is reached. Thus, it follows the dynamics $S_i \rightarrow \text{sgn}\,[h_i(t)]$ where h_i is the local field of the connection of the neurons. The sum of the field induced by each neuron is given as follows.

$$h_i = \sum_k^N \sum_j^N T_{ijk} S_j S_k + \sum_j^N T_{ij} S_j + T_i \tag{13}$$

The task of the local field is to evaluate the final state of neurons and generate all the possible RAN-SAT-induced logic that was obtained from the final state of neurons. One of the most prominent features of the HNN network is the fact that it always converges to stable states (Hopfield, 1982). The Lyapunov energy function (LEF) utilized in HNN for RANkSAT logic programming is presented as follows

$$H_F = \ldots -\frac{1}{3} \sum_{i=1}^N \sum_{j \neq k}^N \sum_{k=1, i \neq k}^N T^{(3)}_{ijk} S_i S_j S_k \ -\frac{1}{2} \sum_{i=1, i \neq j}^m \sum_{j=1}^m T^{(2)}_{ij} S_i S_j - \sum_{i=1, i \neq j}^m T^{(1)}_{i} S_j \tag{14}$$

The energy function of the HNN model is especially critical since it will decide the interoperability of the network. The value obtained from the equation will be verified as global or otherwise. The network would generate the right response when the induced neuron state reached global minimum energy. There are minimal works to integrate HNN with RANkSAT as a single computational network.

2.3 Random k satisfiability reverse analysis (RANkSATRA)

The central emphasis of logic mining is to extract useful logical rules from the data sets provided. One of the extraordinary fields of data mining is logic mining proposed by [22]. It was shown that the information can be expressed in logical form. As a logical rule, the reverse analysis approach (RA) was implemented to derive useful knowledge from real-life data based on the horn clause [22]. In this study, Random k Satisfiability enhanced reverse analysis method or abbreviated as (RANkSATRA) is proposed to extract the optimum RANkSATRA logical rule to explain the behavior of the COVID-19 data sets. In this context, RANkSATRA is a logic extraction method that uses the HNN-RANkSAT model structure to extract from the data set the valuable logical rule (COVID-19 data set). Because of its non-systematic behavior, the RANkSAT logical rule would be used to describe and chart the data sets due to flexibility and convenience. In particular, the RANkSATRA approach can derive the ideal logic representing the relationship between the actual data set attributes of the COVID-19. Pursuing that, to be used in classification or estimation, the secret information in the data set is retrieved.

In our hybrid HNN model, RANkSATRA is performed out to represent a data classification framework in data mining. Inside the RANkSAT clauses, each of the attributes can be translated into atoms. To construct the RANkSAT logical rule, seven attributes from the data sets are then chosen by considering $k \leq 3$. Logic mining is a method that extracts information from a data set using logic programming. In this regard, this section will clarify how our HNN-RANkSATRA model implements the logic mining technique called the random 3-satisfaction-based reverse analysis process (RAN3SATRA) to obtain the relationship of COVID-19 data entries. By acquiring the synaptic weight between 3 neurons, RAN3SATRA might be able to reveal the level of their connectedness.

Consider n attributes of the COVID-19 data sets $(S_1, S_2, S_3, \ldots, S_n)$, where $S_i \in [1, -1]$. All entries are represented in bipolar states. Since this chapter considers $F_{RAN3SAT}$, the arrangement of each S_m consists of S_i, S_j, S_k where $i \neq j \neq k$. For S_m those leads $P_{RANkSAT learn} = 1$, we assign

$$S_m = \left(S_i^{\max[n(s_i)]} \vee S_j^{\max\left[n\left(s_j\right)\right]} \vee S_k^{\max[n(s_k)]} \right) \tag{15}$$

$$S_i = \begin{cases} S_i, & S_i = 1 \\ \neg S_i, & S_i = -1 \end{cases} \tag{16}$$

Based on the obtained S_m, we can formulate $P_{RANkSAT best}$:

$$P_{RANkSAT learn} = \vee_{m=1}^{k} S_m \tag{17}$$

For example, we will choose $G_1 = (S_1 \vee \neg S_2 \vee \neg S_3)$ if $S_1^{\max[n(S_1)]} = S_1$, $S_2^{\max[n(S_2)]} = \neg S_2$, and $S_3^{\max[n(S_3)]} = \neg S_3$. $P_{RANkSAT best}$ will be embedded into HNN. Henceforth, we will obtain the states of S_i that correspond to $E_{F_{RAN3SAT_{best}}} \to 0$. By comparing Eq. (3)

with Eq. (14), the corresponding T_{ijk} will be obtained. During the testing phase, the induced states, S_i^B, will be obtained by obeying Eq. (13). Subsequently, the induced logic P_i^B will be constructed according to logical rule given in Eq. (2). Finally, the chosen induced logic obeys $P_i^B = P_i^{test}$ (Training data). **Figure 1** demonstrates how RANkSATRA has been implemented in the HNN model to classify COVID-19 surveillance data set (CSDS). In this study, RANkSATRA, is used to determine the relationship among the data set. In learning COVID-19 surveillance data set, {detected, not detected} would be converted into bipolar representation {1,-1}, respectively. Each objective taken would be represented in terms of neurons in RANkSATRA. Hence, there would be a total of seven neurons being considered in this data set. Each neuron will be represented with entries COVID-19 surveillance data set.

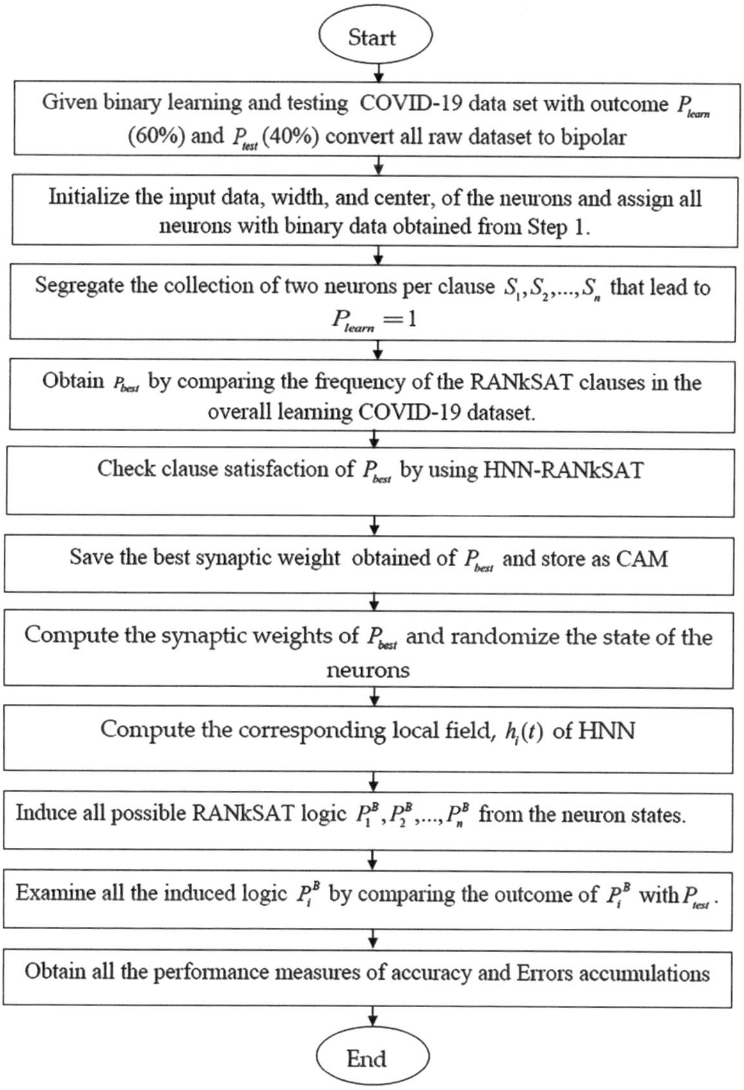

Figure 1.
Implementation of RANkSATRA for COVID-19 surveillance data set.

2.4 RANkSATRA experimental setup

The simulation has been developed to explore the capacity of the Random kSatisfiability reverse analysis (RANkSATRA) in Hopfield neural network for COVID-19 surveillance data set classification. Sixty percent (60%) of the data points in the databases were used for learning data collected, and four ty percent (40%) were used for testing. Microsoft Visual C++ applications running on Windows 8.1, 64-bit, 4.40 GHz CPU, 4GB RAM, and 400 GB hard drive specifications were adopted. For both learning and testing, the overall CPU time is 24 hr. If the model crosses the recommended processor time threshold, the recommended algorithm structure cannot train HNN-based RANkSATRA using real life. In terms of the logical rule that will be embedded inside HNN, the existing work of Sathasivam and Abdullah [29] that implemented HORNSAT in their proposed reverse analysis method. RANMAXkSAT has been proposed [30] and RANkSAT has been proposed in [31]. Both of these proposed models were considered the only existing logic mining in the literature.

2.5 Implementation of COVID-19 surveillance data set

In this chapter, COVID-19 surveillance data set was occupied in RAN*k*SATRA, HORN*k*SATRA, and RANMAX*k*SATRA for classification COVID-19 data set collected from UCI machine learning repository. It contains information about the data set and contains different purposes. The original data contains 7 instances with nine attributes with two classes. The classes are detected and not detected. But, to find out more relevant features from a COVID-19 data set, feature selection methods are applied to the data set. In this experiment, our aim is to that utilize the same data set. The details of COVID-19 are shown in **Table 1**.

3. Performance evaluation metrics

In this section, simulation experiments were performed to assess the performance of the proposed logical rule model on a different number of clauses. The performance evaluation indicators are deployed to analyze the effectiveness of our SATRA model in extracting important logical rule for CSDS. The metrics used in this study measure the performance of the training phase of HNN models. The metric solely indicates the performance of the retrieved neuron state that contributes to the best CSDS classification. During the learning phase, the performance of the RAN*k*SAT representation that governs the network will be evaluated based on the following fitness equation:

$$f_k = \sum_{k=1}^{NC} C_k \tag{18}$$

NC is the number of clauses for any given P_k^B . According to Eq. (18) C_k is defined as follows.

$$C_k = \begin{cases} 1 & True \\ 0 & False \end{cases} \tag{19}$$

Details of each attribute		Output $F_{RANkSAT}$
T1	Fever or history of fever	
T2	signs and symptoms of respiratory distress e.g., cough, cold, and sore throat	
T3	Acute and/or severe pneumonia or Respiratory Infections (ARI)	
T4	There are no other causes based on convincing clinical descriptions	
T5	In the last 14 days before the symptoms have a history of travel or living abroad who reported local transmission	To classify between $F_{RANkSAT} = 1$(Detected) and $F_{RANkSAT} = -1$ (not detected)
T6	In the last 14 days before the symptoms have a history of travel or stay in the local transmission area in Indonesia	
T7	Contact with Coronavirus Disease 2019 (COVID-19) confirmation cases in the last 14 days before symptoms	

Table 1.
List of attributes for COVID-19 surveillance data set (CSDS) [32].

The measurements are evaluated based on the accuracy, and errors accumulation that reflects the network complexity based on the number of neurons using the following formula.

$$TRANING\_MAE = \sum_{i=1}^{n} \frac{1}{n} \left| f_{\max} - f_k \right| \tag{20}$$

$$TRAINMING\_RMSE = \sum_{i=1}^{n} \sqrt{\frac{1}{n} \left(f_{\max} - f_k \right)^2} \tag{21}$$

where $f_{\max}$ and f_k are the output value and target output value, respectively, and n is a number of the iterations.

$$\text{TRAINING\_BIC} = nIn(MSE) + pa.In(n) \tag{22}$$

where n, pa, and MSE indicate the number of solutions obtained, their parameters, and the mean square error used in the model, respectively. Since the HNN is free from any pa [33], the equation is re-written as follows.

$$\text{TRAINING\_BIC} = nIn(MSE) \tag{23}$$

where the MSE is measured in calculating BIC and n depicts the number of iterations during the simulation. Hence, the formula of MSE is given as follows,

$$TRAINING\_MSE = \frac{1}{n} \sum_{i=1}^{n} \left(f_{\max} - f_k \right)^2 \tag{24}$$

Logical rule	MAE	RMSE	ACCURACY
RAN*k*SATRA	1.6140	0.466	96.1
RANMAX*k*STARA	3.921	1.170	83.3
HORN*k*SATRA	2.145	0.845	88.5

Table 2.
Training error and accuracy for all HNN models.

where f_i and f_{max} describe the fitness value observed during each of the execution and the maximum fitness, respectively.

$$TRAINMING\_ACCURACY = \frac{P_{induced}^{Correct}}{N_{P_{test}}} \times 100\% \quad (25)$$

The performance of the HNN on the proposed logical rule is presented in **Table 2**.

4. Result and discussion

The performance of the simulated program with different complexities for neurons in the HNN-RAN*k*SATRA model will be evaluated with the existing models HNN-HORNSAT [29] and RANMAX*k*SAT [34] in terms of root mean square error (RMSE), mean absolute error (MAE), Bayesian information criterion (BIC), accuracy, and CPU time. **Figures 2** and **3** illustrate the root mean square error (RMSE) and mean absolute error (MAE) of HNN models during the training process. This analysis only considers $1 \leq NC \leq 10$. The COVID-19 surveillance data set (CSDS) data have successfully been embedded into the network and forming variants of learnable Boolean k Satsifiability logic, RANkSAT with the existing HORNSAT and RANMAX*k*SAT. A comparison has been made between the variants of satisfiability logic for COVID-19 surveillance data set (CSDS) data classification. As seen in

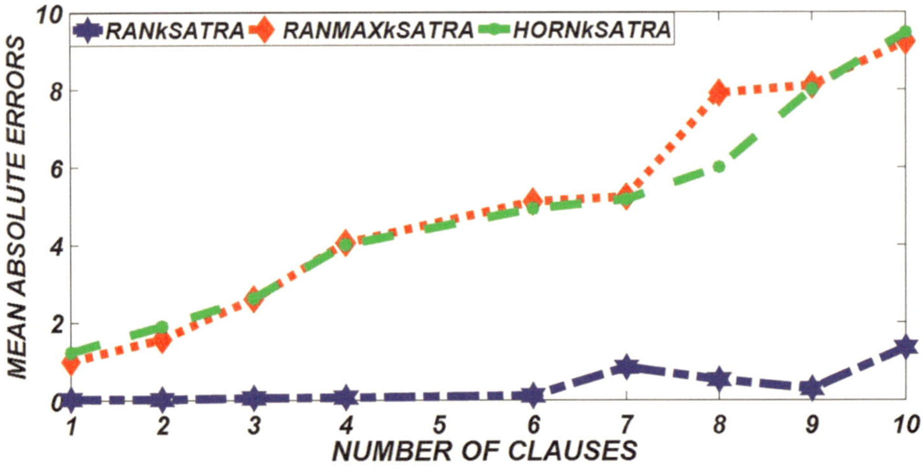

Figure 2.
MAE evaluation of HNN models for COVID-19 classification.

Figure 2, HNN-RAN*k*SATRA with NC = 1 until NC = 10 has the best performance in terms of RMSE compared to HNN-HORN*k*SATRA and RANMAX*k*SATRA. This is because the HNN-RAN*k*SAT utilizes random logical inconsistencies to derive the optimum synaptic weight for HNN. Optimal synaptic weight is a building block for optimum CSDS classification. The RMSE result from **Figure 2** has been supported by the value of MAE in **Figure 3**.

The investigation of a model's performance is separated into two parts. The first significant part is to examine the quality of the solution generated by different searching techniques by employing suitable training errors. Secondly is to analyze the robustness and efficiency of the proposed model by comparing *CT* and *Q* needed to execute the models' mechanism. There are five performance evaluation metrics involved to analyze the training and testing stage of our modified models as presented in the performance evaluation section. Therefore, this research's main contribution is to portray the competency of HNN in SAT in outperforming the existing models.

In **Figure 2**, HNN-RAN*k*SATRA with NC = 2 has the best performance in terms of MAE. It can be observed that MAE for NC = 2 is equal to 0.027 compared with NC = 10 which recorded 1.3515. The searching process of HNN for HORN*k*STRA and RANMAXkSTRA displayed a similar error trend with RANMAXkSTRA has the highest error at NC = 8. In **Figure 3**, HNN-RAN*k*SATRA has the best performance in terms of RMSE. It can be observed at NC = 1, the RMSE is equal to 0.0323 and 1.3995 was recorded at NC = 10, which are lower than HNN-HORNSAT and HNN-RANMAX-*k*STRA. The HNN-RANMAX*k*STRA recorded the highest RMSE with 0.832 at NC = 1 and 10.4602 at NC = 10. The searching process of HNN for RAN*k*STRA and HORN*k*SATRA displayed a similar RMSE trend with closed merging. As the number of neurons increased, the learning phase of HNN models was convoluted because the network required to find the consistent interpretation for the best logic for optimal COVID-19 surveillance data set (CSDS) classification. In this case, the learning phase of HNN for both RANMAX*k*STRA and Horn*k*SATRA was trapped in the trial and error search process, which resulted in high RMSE and MAE accumulations. However, the RMSE and MAE recorded by HNN-HORNSAT and RANMAX*k*STRA were to some extent higher than the RAN*k*STRA. At *NC* = 5, the value of MAE and RMSE are approximately 57% times bigger than *NC* = 10 because for the network to converge into

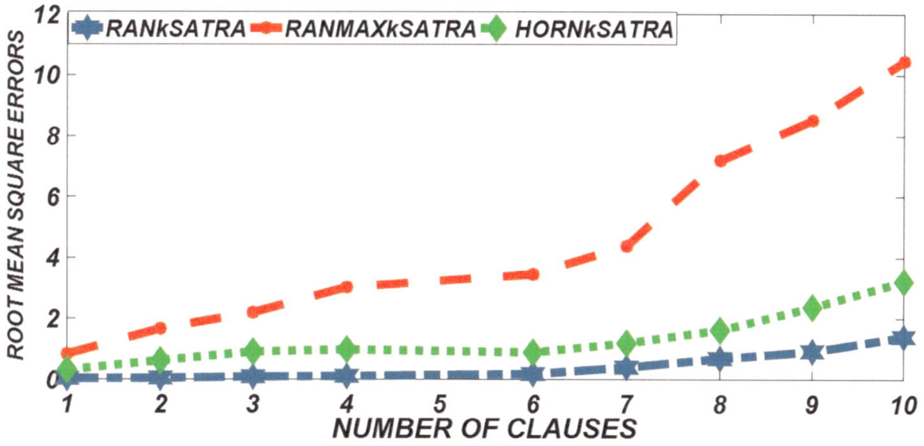

Figure 3.
RMSE evaluation of HNN models for COVID-19 classification.

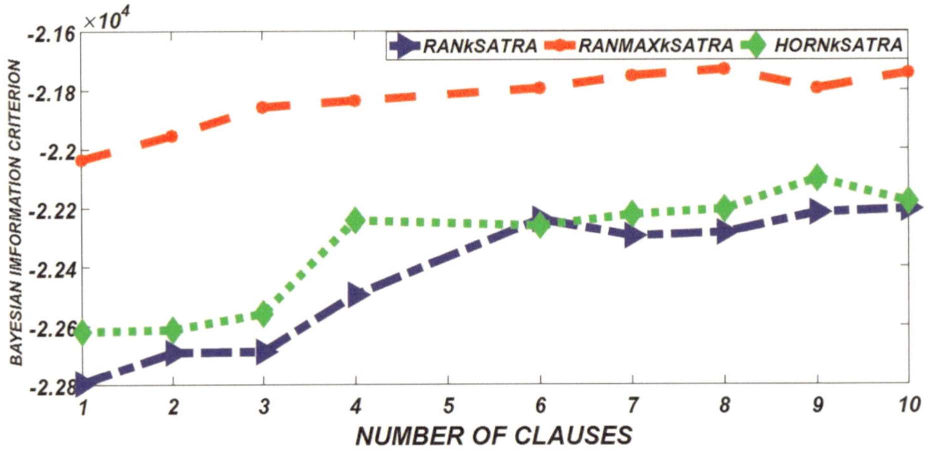

Figure 4.
BIC evaluation of HNN models for COVID-19 classification.

full fitness (learning completed), more iterations are needed. Therefore, a similar trend can be seen in the HNN model as the complexity increases. Thus, HNN model works optimally in learning the variants of satisfiability logic entrenched to the network before being stored into content addressable memory. The complete learning process will ensure the network generates the best logic to represent the characteristic of the HNN for optimal COVID-19 surveillance data set (CSDS) classification.

In contrast, the learning phase in HNN-HORNSAT and HNN- RANMAX*k*STRA were computationally expensive as more iterations were needed leading to higher RMSE and MAE values compared to HNN-RAN*k*SATRA. All in all, SATRA contributes to generating the best logic to represent the relationship between each instance and the verdict of HNN for COVID-19 surveillance data set (CSDS) classification. The MAE and RMSE results displayed in **Figures 2** and **3** were supported by BIC in **Figure 4**. The "best" model will be the one that neither under-fits nor over-fits. Although the BIC will choose the best model from a set of models, it will not say anything about the absolute quality of the model. However, the HNN-RAN*k*SATRA is the best choice for COVID-19 surveillance data set (CSDS) classification. In terms of BIC, HNN-RAN*k*SATRA outclasses other models. The accumulation of MSE tends to penalize the values of BIC. The *BIC* for HNN-RAN*k*SATRA is, therefore, the lowest compared to the other two models.

Table 3 displays the CPU Time results for the HNN models, respectively. To assess the robustness of the models in logic mining, CPU time is recorded for the learning and retrieval phase of HNN. HNN less CPU time is required to complete one execution of learning and testing for CSDS classification when the number of NC deployed is less. As it stands, HNN-RAN*k*SATRA model, when the complexity is higher, models take a long time to finish the learning process. Overall, the HNN continues to be proficient at reducing the k Satisfiability inconsistencies and computing the global solution in a reasonable amount of time on the CPU. Because there are more instances to handle during the learning and testing phase of HNN, the CPU Time for HNN-RANMAXkSAT is consistently greater than HNN-RANkSATRA and HNN-HORNSATRA. However, the CPU time recorded for the existing methods was higher due to more iterations needed in generating the best logic for the HNN.

NC	RAN*k* SATRA	RANMAX*k* SATRA	HORN SATRA
1	0.643	1.8232	1.735
2	0.835	2.0279	2.8006
3	1.9707	5.3258	6.1323
4	3.2358	11.1566	13.004
6	9.7422	19.434	24.9292
7	18.46	22.5552	35.1675
8	25.2815	36.9081	42.005
9	37.958	65.6528	58.0088
10	72.312	91.41	76.234

Table 3.
Computation (CPU) time(s) for the HNN model.

Table 2 displays the testing error and accuracy data collected for each model when the HNN was tested. As a result, for each of NC = 1 through NC = 8, the testing RMSE, MAE, and accuracy recorded for HNN-RANkSATRA, HNN-RANMAXkSAT, and HNN-HORNSAT were consistently identical. The ability of our suggested network, SATRA and HNN-kSAT, to produce the optimal logic during the learning phase, which contributes to a very low error, is thus demonstrated by this. According to performance evaluation metrics captured during simulation, the learning mechanism in SATRA is capable of deriving the best logic to map the relationships between the characteristics in HNN. 96.30 percent accuracy was attained by the suggested model, as measured by the accuracy that each model recorded. The existing works on logic mining in HNN can still achieve up to 80% of the CSDS classification accuracy. To sum up, HNN-RAN*k*SATRA is the best model for learning and testing HNN due to lower values of RMSE, MAE, and the highest accuracy for a logical rule. Hence, the logical rule obtained from HNN-RAN*k*SATRA will benefit the healthcare sector in the CSDS classification problem.

5. Conclusion

In this research, the formulation of constructing HNN proved to be adequate to represent the mechanism in RAN*k*SAT the implementation of EA with standalone HNN framework with reverse analysis proved to be effective in solving real-life data sets. In order to optimally express the relationship in a data set, we successfully turned the COVID-19 data set into an ideal logical representation in the form of RANkSAT representation in this chapter. Furthermore, we used COVID-19-RANkSATRA as a substitute method for identifying correlations between the variables that correspond to classification between $F_{RANkSAT} = 1$ (Detected) and $F_{RANkSAT} = -1$ (not detected) of the COVID-19 data set. To work, the neural network needed to be trained. Because the structure of a neural network (NN) differs from the structure of microprocessors, it must be imitated. For large neural networks, it required a lot of processing time and space in weight training and adjustment, system adaptation for determining the number of layers, node transfer functions, retrieval phase, and learning rules.

One of the limitations of Hopfield type of artificial neural network is that it sometimes gets trapped at the local solution (premature convergence) instead of the global one. Our Future direction is to incorporate novel metaheuristics algorithms such as Election algorithm, genetics algorithm, dragonfly algorithm, etc. to enhance the performance of Hopfield type of artificial network for better training and retrieval process in preventing the Hopfield type of artificial network in settling in the local solution for better searching and classification problems. We will further utilize other variants of Boolean satisfiability such as k-SAT, Rondom Half-SAT, MAX-kSAT, Random NAE-SAT, and XOR-SAT for better optimization problems. Various type of data set, such as agriculture, financial, actuarial as well as environmental data set will also be used in our future studies.

Author details

Hamza Abubakar[1,2]* and Muntari Idris[2]

1 University of Science, School of Mathematical Sciences, Malaysia

2 Department of Mathematics, Isa Kaita College of Education, Dutsin-ma, Katsina State, Nigeria

*Address all correspondence to: zeeham4u2c@ikcoe.edu.ng

References

[1] Roberts M, Driggs D, Thorpe M, Gilbey J, Yeung M, Ursprung S, et al. Common pitfalls and recommendations for using machine learning to detect and prognosticate for COVID-19 using chest radiographs and CT scans. Nature Machine Intelligence. 2021:**3**(3);199-217

[2] Chiroma H, Ezugwu AE, Jauro F, Al-Garadi MA, Abdullahi IN, Shuib L. Early survey with bibliometric analysis on machine learning approaches in controlling COVID-19 outbreaks. PeerJournal of Computer Science. 2020; **6**:e313

[3] Elaziz MA, Hosny KM, Salah A, Darwish MM, Lu S, Sahlol AT. New machine learning method for imagebased diagnosis of COVID-19. PLoS One. 2020;**15**(6):e0235187

[4] Manap HH, Md Tahir N, Yassin AIM. Statistical analysis of Parkinson disease gait classification using artificial neural network. IEEE International Symposium on Signal Processing and Information Technology ISSPIT 2011. 2011. pp. 060-065

[5] Grover S, Bhartia S, Akshama YA, Seeja KR. Predicting severity of Parkinson's disease using deep learning. Procedia Computer Science. 2018;**132**: 1788-1794

[6] Ström F, Koker R. A parallel neural network approach to prediction of Parkinson's disease. Expert Systems with Applications. 2011;**38**(10): 12470-12474

[7] Bind S, Tiwari AK, Sahani AK. A survey of machine learning based approaches for Parkinson disease prediction. International Journal of Computer Science Information Technology. 2015;**6**(2):1648-1655

[8] Saritas I. Prediction of breast cancer using artificial neural networks. Journal of Medical Systems. 2012;**36**(5):2901-2907

[9] Karayilan T, Kiliç Ö. Prediction of heart disease using neural network. In: 2017 International Conference on Computer Science Engineering (UBMK). 2017. pp. 719-723

[10] Pan S, Warwick K, Stein J, Gasson MN, Wang SY, Aziz TZ, et al. Prediction of Parkinson's disease tremor onset using artificial neural networks. Proceedings of the fifth IASTED International Conference: biomedical engineering 2007. pp. 341-345

[11] Wu D, Warwick K, Ma Z, Gasson MN, Burgess JG, Pan S, et al. Prediction of parkinson's disease tremor onset using a radial basis function neural network based on particle swarm optimization. International Journal of Neural Systems. 2010:**37**(4);2923-2928

[12] Vijayarani S, Dhayanand S, Professor A. Research scholar MP. Kidney disease prediction using SVM and ANN algorithms. International Journal of Computing and Business Research ISSN (Online). 2015;**6**(2):1-12

[13] Kasikumar K, Mohamed Najumuddeen M, Suresh R. Applications of data mining techniques in healthcare and prediction of heart attacks. International Journal of Data Minimum Technology Application. 2018;**2**(02): 250-255

[14] Tang J, Yang B, Adams MP, Shenkov NN, Klyuzhin IS, Fotouhi S, et al. Artificial neural network–based prediction of outcome in Parkinson's disease patients using DaTscan SPECT imaging features. Molecular Imaging and Biology. 2019;**21**(6):1165-1173

[15] Ibrahim F, Taib MN, Abas WABW, Guan CC, Sulaiman S. A novel dengue fever (DF) and dengue haemorrhagic fever (DHF) analysis using artificial neural network (ANN). Computer Methods and Programs in Biomedicine. 2005;**79**(3):273-281

[16] Ardi Handojoseno AM, Shine JM, Nguyen TN, Tran Y, Lewis SJG, Nguyen HT. Analysis and prediction of the freezing of gait using EEG brain dynamics. IEEE Transactions on Neural Systems and Rehabilitation Engineering. 2015;**23**(5):887-896

[17] Wang X, Zhang M, Zhu J, Geng S. Spectral prediction of Phytophthora infestans infection on tomatoes using artificial neural network (ANN). International Journal of Remote Sensing. 2008;**29**(6):1693-1706

[18] Shantakumar BP, Kumaraswamy YS. Intelligent and effective heart attack prediction system using data mining and artificial neural network. European Journal of Scientific Research. 2009;**31** (4):642-656

[19] Khan MM, Mendes A, Chalup SK. Evolutionary wavelet neural network ensembles for breast cancer and Parkinson's disease prediction. PLoS One. 2018;**13**(2):e0192192

[20] Rau HH, Hsu CY, Lin YA, Atique S, Fuad A, Wei LM, et al. Development of a web-based liver cancer prediction model for type II diabetes patients by using an artificial neural network. Computer Methods and Programs in Biomedicine. 2016;**119**(1):29-42

[21] Janghel RR, Shukla A, Tiwari R, Kala R. Breast cancer diagnosis using artificial neural network models. The 3rd International Conference on Information Sciences and Interaction Sciences ICIS. 2010. pp. 89-94

[22] Sathasivam S, Wan Abdullah WAT. Logic mining in neural network: Reverse analysis method. Computing (Vienna/ New York). 2011;**91**(2):119-133

[23] Alzaeemi SAS, Sathasivam S. Examining the forecasting movement of palm oil Price using RBFNN-2SATRA metaheuristic algorithms for logic mining. IEEE Access. 2021;**9**:22542-22557

[24] Alzaeemi SA, Sathasivam S. Artificial immune system in doing 2-satisfiability based reverse analysis method via a radial basis function neural network. PRO. 2020;**8**(10):1295

[25] Sathasivam S, Mansor MA, Kasihmuddin MSM, Abubakar H. Election algorithm for random k satisfiability in the Hopfield neural network. PRO. 2020;**8**(5):568

[26] Uykan Z. On the working principle of the Hopfield neural networks and its equivalence to the GADIA in optimization. IEEE Transactions on Neural Networks and Learning Systems. 2020;**31**(9):3294-3304

[27] Sathasivam S. Enhancing logic programming performance in recurrent Hopfield network. European Journal of Scientific Research. 2009;**37**(1):1-7

[28] Bukov M, Day AGR, Sels D, Weinberg P, Polkovnikov A, Mehta P. Reinforcement learning in different phases of quantum control. Physical Review X. 2018;**8**(3):031086

[29] Sathasivam S, Abdullah WAT. Logic mining in neural network: Reverse analysis method. Computing. 2011;**91** (2):119-133

[30] Abubakar H. Neuro-symbolic integration of Hopfied neural network for optimal maximum random kSatisfiability (MAXRkSAT)

representation. Journal of Reliability and Statistical Studies. 2020:199-220

[31] Abubakar H, Sathasivam S. Developing random satisfiability logic programming in Hopfield neural network. AIP Conference Proceedings. 2020;**2266**(1):040001

[32] Kemenkes RI. Pedoman Pencegahan dan Pengendalian Coronavirus Disease (COVID-19). Germas. 2020;1(2):136-151

[33] Hamadneh N, Sathasivam S, Tilahun SL, Choon OH. Learning logic programming in radial basis function network via genetic algorithm. Journal of Applied Sciences. 2012;**12**(9):840-847

[34] Abubakar H, Abdu Masanawa S, Yusuf S. Neuro-symbolic integration of Hopfield neural network for optimal maximum random kSatisfiability (Maxrksat) representation. Journal of Reliability and Statistical Studies. 2020; **49**(2)

Chapter 4

Industrial Fluids Components Health Management Using Deep Learning

Vidyadevi G. Biradar, H.C. Nagaraj, S.G. Mohan and Piyush Kumar Pareek

Abstract

The fatigue state of fluid components such as valves, metal surfaces in gas or oil carrying pipelines is important to monitor on regular basis and plan for repair work to avoid risks associated with them, this becomes more crucial when the pipelines are supplying hazard prone fluids. There exist methods for detection of corroded surfaces, scratches and fractures in pipelines, valves, and regulators etcetera. The conventional methods are based on sensors and chemical analysis methods. There are challenges with conventional methods pertaining to the desired metric of scalability and disadvantages of these methods is they are contact based and destructive methods. Therefore, to overcome these limitations of existing methods there is a need for development of non-contact and nondestructive methods. The recent advancements in Artificial Intelligence technology in every domain including health care monitoring, agriculture sector, defense applications and civilian applications etc., have shown that deep learning methods can be explored in industrial applications to develop fault tolerant systems which help fluid components state of health monitoring through computer vision. In this chapter proposes various methods for analysis of health state of fluid components using deep convolutional neural networks and suggest the best models for these applications.

Keywords: deep learning, fluid component, convolutional neural network

1. Introduction

In oil and gas transportation industry metal pipelines are the major transport means for transporting fluids such as crude oils, petrochemical products for long distance. Due to various environmental conditions and fatigue induced in pipelines as a result of fluid pressure over a period time of operation the health of pipelines will be deteriorated due to corrosion, dents, and damages and other reasons which leads to various types of pipelines defects [1]. Therefore, timely maintenance of pipelines plays vital role in avoidance of untoward incidents and economic loss etc. the early detection of pipeline defect helps in planning preventive maintenance to mitigate corrosion progression, obstacle in flow etcetera and thus reduce maintenance costs [2, 3]. The conventional methods such as crack detection using camera, magnetic field, acoustic methods, and thermal camera are very useful methods and demonstrate satisfactory performance [4], however, these are tedious and dependable on environmental conditions. Therefore, there is a need for automatic techniques. Typical pipeline defects are caused due to metal loss, dents, stress induced cracks, gouge, and coating damage etc. The quick and reliable detection of leakages is very much essential to avoid hazards. It is required to ensure that these fluids are safely transported to the destination.

Identifying leaks at right time is essential to avoid serious problems. Methods used for leakage detection are, i) Distributed temperature monitoring approaches utilizing optical fibers to identify and localize leakages, ii) acoustic impact monitoring method, iii) artificial neural network technique and, Leakage detection techniques needs improvement to achieve greater precision in identification of defect location [3].

The objectives of books chapter are listed below.

1. To understand the importance of fluid components health

2. Deep learning models

3. Pre-trained convolutional neural network models for fluid components health monitoring

4. Transfer learning

5. Conclusion and future scope.

The first objective of the chapter is discussed in introduction section, Section 2 gives insight into deep learning methods for fluid component damage detection and design of convolutional neural networks, Section 4 gives guidelines for transfer learning strategies and Section 5 presents conclusion.

This chapter provides insight into the alternate method for pipeline damage detection is deep learning paradigm. This chapter presents practical perspectives of convolutional neural network and provides guidance on transfer techniques to tune the pretrained model to solve the problem.

2. Deep learning models for pipeline damage detection

In industries defect detection is performed by an expert human expert to analyze the defect patterns [5], manual analysis by experts is tedious and time-consuming job, therefore, there is a need for automated technique. The automated techniques which utilize computer vision help to solve these challenges.

The identification and classification of pipe damages such as cracks with image analysis combined with neuro fuzzy algorithms are presented in [6], here the significant features considered for characterization were features from Hough transforms, morphological operations, shape statistics, regression analysis and eigen vector analysis. The defects are classified using back propagation algorithm.

Artificial neural networks (ANNs) possess learning skills and capable of adapting themselves to alterations in the training phase. ANNs are interrelated groups of neurons, neural networks are used in modeling complex connections between inputs and outputs, neural networks gave good results in detection of cracks in pipelines [3].

Pipe cracks detection using computer vision helps to solve the issues of conventional methods, however, with complex background content in the image makes problem solving a challenging job. A method based on computer vision is presented in [7]. The methodology of detecting cracks apply image filtering for background subtraction using tuned adaptive thresholding technique and crack contour is extracted using morphological operator. To understand the depth of the crack 3D visualization is performed using successive images [8].

Pipeline cracks are detected and classified using image processing techniques, a method which converts RGB into gray scale image, apply Sobel operator to edge points extraction and through edge linking judge the artifacts are holes, cracks etc., is presented in [5].

Deep learning models provide flexibility in the process of defect detection as the model adapts to the dataset in learning relevant features and therefore give higher success rate of defect identification and classification. The purpose of defect detection depends on the application. In some cases, just detecting the presence of a defect is sufficient and others, classification and labelling are important [3, 9].

Image processing and machine learning algorithms largely depend on the accuracy of image feature extraction which is a challenging job and which intern depends on the image quality. Therefore, automatic feature extraction using data driven methods come for rescue. The convolutional neural networks are widely used for automatic feature extraction and classification. The deep learning models have shown highest success among the state-of-the-art approaches to solve computer vision problems [10]. A deep convolutional neural network with configurations of convolutional layers, filters, batch normalization and pooling in different combination are experimented for determining optimal hyperparameters of the model and encouraging results are obtained.

3. Convolutional neural networks

The convolutional neural networks (CNN) consist of various types of layers with specific functionality. The convolutional layer convolves input image with various filters whose co-efficient are determined during the run of backpropagation algorithm in number of iterations. The convolution operations are carried out using a set of filters to extract images features. The CNN may have any number of convolutional layers as shown in the **Figure 1**, in this, initial layers extract primitive level features from input image such as edges whereas later layers extract high levels features such as shape features and contour etc. The pooling layers reduces feature dimension by extracting significant pixels and down sampling images, min, max, avg. pooling are the typical available options. The drop out layer dynamically changes the pool of neurons in the network to improve the generalization of the model to avoid overfitting, fully connected layer classify the images based on the features extracted by the series of convolution layers [11]. To improve the speed of the model training and avoid vanishing gradient problem activation layer are added to CNN network, different types of activation functions such as Tanh, ReLu, Sigmoid and Softmax are used. Among these the popular one is ReLu and its variants [12].

An input layer, hidden layers, and output layer are different types of layers in CNN. The design of the neural networks is the way human brain works. Input layers

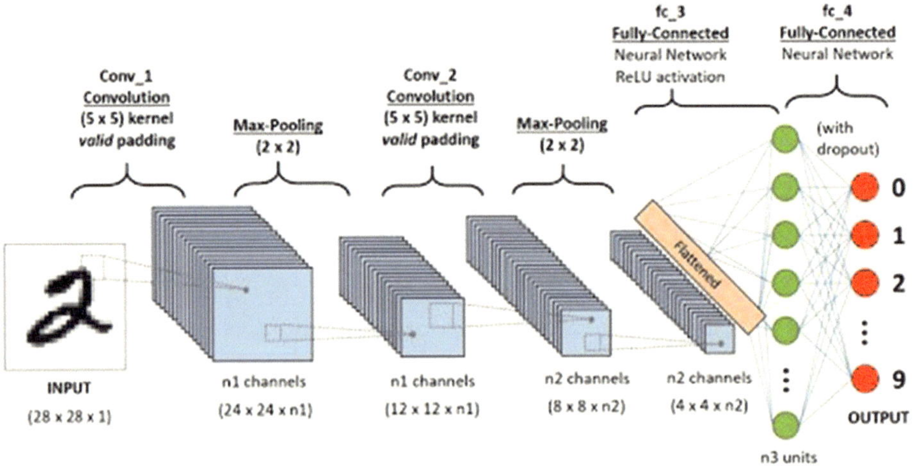

Figure 1.
Architectural components of CNN.

collect inputs, process them, and produces the result. CNN has many hidden layers which perform convolution operations to extract features from the input image. The features are classified by the fully connected layer. There are different types of CNN models, all types have convolutional and pooling layers.

Figure 2 depicts convolution operation, in this operation a filter or kernel is convolved with the input image. The next layer takes in the output generated by first layer and so on. Convolutions operation in image processing are applied to sharpen, blur images and edge detection etc. CNNs establish a connection pattern between neurons of adjacent layers with drop out technique.

The layer of CNN generates number of activation maps from the input image, and these are then fed to the subsequent layers, this process in shown in **Figure 3**. The primitive features such as horizontal and vertical edges are most of the times extracted in the earlier layers. The later layers extract high level features like objects, shape of the object and features which helps in making sense of features.

Pooling Layer – the dimension of feature map is reduced by pooling layers. This in turn reduces the number of parameters of the network and computation time.

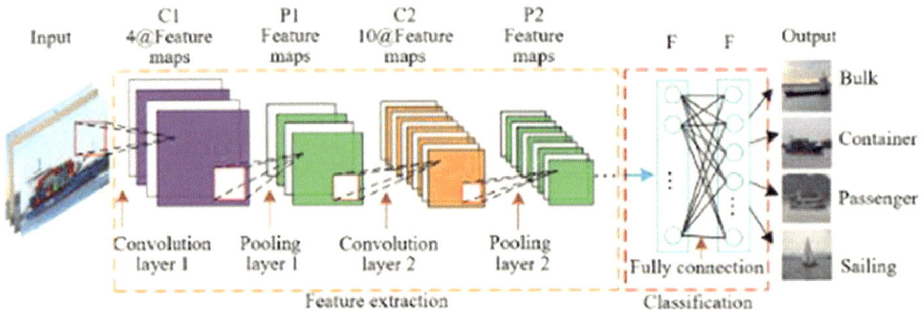

Figure 2.
Convolution operation.

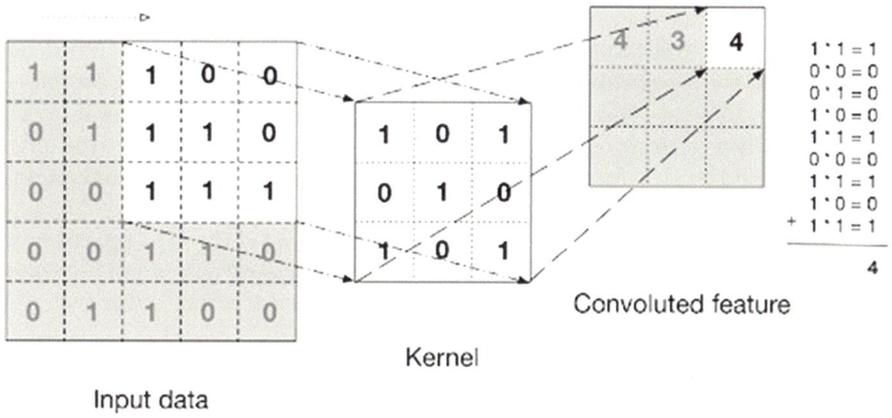

Figure 3.
Feature extraction in CNN.

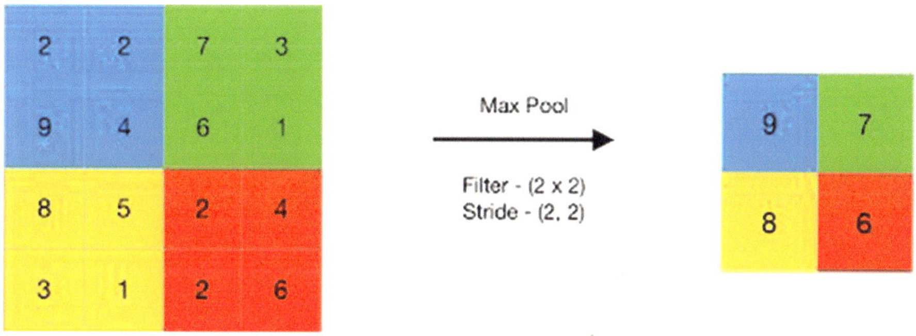

Figure 4.
Max pooling operation.

The down sampling of image using max pooling operation is shown in the **Figure 4** it selects max pixel value from 2x2 image patch and uses this single value to represent feature in the image. In a similar manner, min pooling and average pooling works where in these choose min and average pixel values respectively.

4. Pretrained models for pipeline health detection

This section gives an insight into detailed methodology which is suitable for fluid components health detection using deep convolutional neural networks and covers image augmentation techniques.

The deep learning models automatically determining features are of interest for pipelines defect classification based on the knowledge acquired during training of huge dataset. To understand the behavior of the AI model, model interpretation tools such as Heat maps and Grad CAM tools are used. A U-Net architecture-based model is presented in [13], a method based on YOLOv5 model demonstrated in [14] which uses X-Ray images for detection of pipeline weld defects, the results indicated that YOLOv5 performs better than R-CNN interns of metrics such as speed of detection and classification accuracy.

The detection of cracks which are thin, irregular in shape and with complex texture background in the image makes crack detection task challenging. The deep learning models show improved performance in cracks detection in fluid components, however, there are many bottlenecks for achieving the best performance. These include inadequate training data, imbalanced data and manually labeled data and absence of ground truth information. In addition to this, the computational requirements are very high as deep learning involve training periods [9].

The research work in [15], combines fuzzy digital twin (FDT) and support vector machine (SVM), with backstepping (BS) and good improvement in accuracy compared other methods.

Deep learning tools aid in identification of pipeline damages at the earliest time. This paper utilizes convolutional neural networks for diagnosis of pipeline threats [3].

The CNN model deployed on AUV (autonomous underwater vehicle) is used for detection of underwater pipeline detection, the challenges addressed were distortion in the focus, contrast, and color. The scarcity of the balanced dataset is another issue as number of damaged pipeline sample were less compared to the total number of pipelines. The dataset is preprocessed through augmentation techniques such as flip, scale, shift etc. and generate pipeline sample images which fall in small subset. The issue of color distortion is eliminated by converting images into gray scale, background is eliminated using segmentation. The various models that are experimented for pipeline damage detection are custom architecture CNN, VGG and MobileNet, and among these models, MobileNet outperformed the other models [16].

5. Image augmentation

Additional images are generated from the original by slightly modifying it by adding noise, cropping, changing the contrast and by rotating etc.

The image can be flipped vertically and horizontally, it can be rotated left and right by any angle, the size of the image can be changed, regions of interest can be cropped to generate additional sample images, image can be translated in left, right direction and various types of noises can be added [17].

6. Transfer learning

Transfer learning eliminates the dependency of deep learning models on huge dataset which are required for learning features. The process of transfer learning involves application of a pre-trained model for classification of specific domain image dataset. To customize pretrained model for specific dataset, minor changes to the original architecture can be done and fresh training of convolutional layers can be done. However, the model tuning process is based on trial-and-error method and the model hyperparameters are experimentally determined. A comprehensive survey on transfer learning techniques is given in [18]. A classical example of transfer learning in the context of solving classification of COVID-19 images is discussed in [19].

A typical CNN has two parts, they are convolution layers and classification layers. In transfer learning, the classifier is changed as per the new classification problem. There are different possible ways of finetuning the model, it requires training the model on a new dataset to learn problem specific features.

Hyperparameters of CNN include Learning rate, Momentum, Epochs, Batch size, Filter size, Activation layers, Number of hidden layers, Filter coefficient initialization and Dropout which are very important to achieve the performance of the model. The setting optimal values for these hyperparameters is important to achieve desired level of model accuracy and avoid model overfitting [20, 21].

7. Guidelines for training of pre-trained model for repurposing

Select suitable pretrained model for new problem, the options available are VGG, InceptionV3, and ResNet5, DarkNet and YOLO etc., presented in [22], the guideline is, select pretrained model which has been trained on some medical images in case repurpose to classify another type of medical images. As it discussed in the previous sections in the technique of transfer learning, earlier layers of CNN are freezed and layer set of layers are unfreezed and retrained to learn knowledge from the new dataset. Now, decision must be made on how many layers to freeze and how many to unfreeze and retrain. This depends upon the similarity and size of the present dataset with the dataset on which the model is pre-trained. The guideline is, i) if the size of the present dataset is large in size then reuse the architecture and retrain all the layers of pretrained model, ii) if the dataset is similar to the dataset on which the model is pretrained and the present dataset is large, then just retrain the classifier layers, iii) the present dataset is small in size and different from the dataset used while pretrained model, this is a difficult situation, here the dataset needs to be enlarged through augmentation and by generating synthetic images. The pre-trained model must be trained considering the fact that freshly training deeper layers require high end computational GPPs and time.

8. Conclusion

In oil and gas industry timely metal pipelines damage detection plays important role for planning on maintenance planning which is essential in avoiding hazard and reducing economic loss. The pipelines are regularly inspected for damage detection, the maintenance is carried out by conventional methods with human expert investigations. These methods are time consuming and sometimes are subjective in nature. Therefore, there is need for an automatic tool for pipelines damage detection. There exist several methods which are based on image processing, computer vision and machine learning. However, the correctness of these systems is largely dependent on the extraction of robust features techniques. The feature extraction step is the bottleneck for most of the classification algorithms. In recent years deep learning models have outperformed all other methods, these models automatically extract required features and makes classification task more accurate. The deep convolutional neural networks are very popular in solving classification problems. In this chapter convolutional neural networks are investigated for pipelines damage detection and classification. The contributions of this work include guidelines on design and implementation of convolutional neural networks, and directions are given for carrying out transfer learning to repurpose pretrained models. This chapter provides brief discussion on various methods of pipeline damage detection using convolutional neural networks by conducting survey of state-of-the-art-techniques. This chapter also provides insight into design and working of convolutional neural networks.

Author details

Vidyadevi G. Biradar*, H.C. Nagaraj, S.G. Mohan and Piyush Kumar Pareek
NITTE Meenakshi Institute of Technology, Bengaluru, India

*Address all correspondence to: vidyadevi.g.biradar@nmit.ac.in

References

[1] Al-Janabi YT. An overview of corrosion in oil and gas industry: Upstream, midstream, and downstream sectors. Corrosion Inhibitors in the Oil and Gas Industry. 2020:1-39

[2] Mahmoodzadeh Z et al. Condition-based maintenance with reinforcement learning for dry gas pipeline subject to internal corrosion. Sensors. 2020;**20**(19):5708

[3] Jafari R et al. Deep learning for pipeline damage detection: An overview of the concepts and a survey of the state-of-the-art. In: 2020 IEEE 10th International Conference on Intelligent Systems (IS). Bucharest, Romania: IEEE; 2020

[4] Hashim AS, Grămescu B, Nițu C. Pipe cracks detection methods–a review. In: International Conference of Mechatronics and Cyber-Mixmechatronics. Cham: Springer; 2018

[5] Ashraful AM et al. An algorithm to detect and identify defects of industrial pipes using image processing. In: The 8th International Conference on Software, Knowledge, Information Management and Applications (SKIMA 2014). United Kingdom: IEEE; 2014

[6] Sinha SK, Karray F, Fieguth PW. Underground pipe cracks classification using image analysis and neuro-fuzzy algorithm. In: Proceedings of the 1999 IEEE International Symposium on Intelligent Control Intelligent Systems and Semiotics (Cat. No. 99CH37014). United Kingdom. 1999

[7] Huang L et al. Research on pipe crack detection based on image processing algorithm. International Journal of Wireless and Mobile Computing. 2021;**20**(4):328-335

[8] Iyer S, Sinha SK. A robust approach for automatic detection and segmentation of cracks in underground pipeline images. Image and Vision Computing. 2005;**23**(10):921-933

[9] Datta S, Sarkar S. A review on different pipeline fault detection methods. Journal of Loss Prevention in the Process Industries. 2016;**41**:97-106

[10] Khan SM, Haider SA, Unwala I. A deep learning based classifier for crack detection with robots in underground pipes. In: IEEE 17th International Conference on Smart Communities: Improving Quality of Life Using ICT. Taylor & Francis and Routledge; 2020. pp. 78-81. DOI: 10.1109/HONET50430.2020.9322665

[11] Wu J. Introduction to convolutional neural networks. National Key Lab for Novel Software Technology. 2017;**5**(23):495

[12] Dubey SR, Singh SK, Chaudhuri BB. Activation functions in deep learning: A comprehensive survey and benchmark. Neurocomputing. 2022

[13] Melching D et al. Explainable machine learning for precise fatigue crack tip detection. Scientific Reports. 2022;**12**(1):1-14

[14] Yang D et al. Deep learning-based steel pipe weld defect detection. Applied Artificial Intelligence. 2021;**35**(15):1237-1249

[15] Piltan F, Kim J-M. Crack localization of pipelines using machine learning and fuzzy digital twin. In: 2021 International Conference on Information and Communication Technology Convergence (ICTC). Springeropen, IEEE; 2021

[16] Shi J et al. Automated underwater pipeline damage detection using neural nets. In: ICRA 2019 Workshop on Underwater Robotics Perception. 2019

[17] Shorten C, Khoshgoftaar TM. A survey on image data augmentation for deep learning. Journal of Big Data. 2019;**6**(1):1-48

[18] Zhuang F et al. A comprehensive survey on transfer learning. Proceedings of the IEEE. 2020;**109**(1):43-76

[19] Sahinbas K, Catak FO. Transfer learning-based convolutional neural network for COVID-19 detection with X-ray images. In: Data Science for COVID-19. Elsevier, Academic Press; 2021. pp. 451-466

[20] Hinz T et al. Speeding up the hyperparameter optimization of deep convolutional neural networks. International Journal of Computational Intelligence and Applications. 2018;**17**(02):1850008

[21] Aszemi NM, Dominic PDD. Hyperparameter optimization in convolutional neural network using genetic algorithms. International Journal of Advanced Computer Science and Applications. 2019;**10**:6

[22] Han X et al. Pre-trained models: Past, present and future. AI Open. 2021;**2**:225-250

Chapter 5

Determination of the Elastic Constants of a Metal-Laminated Composite Material Using Artificial Neural Networks

Marta Eraña-Díaz and Mario Acosta-Flores

Abstract

This chapter explores the use of an artificial neural network (ANN) to obtain the elastic constants of the components of a metal laminated composite material (MLCM). The dataset for the training and validation of the ANN was obtained by applying an analytical model developed for the study of stresses in MLCM. The information used in the dataset corresponds to MLCM configurations and data generated with variants registered in the structural presentation of the inputs and outputs. The best configuration found for the generation of the ANN models yielded an average relative error of less than 4% in relation to the results of the constants evaluated and published in a previous article. As shown in this research, it is important to have a clear definition of the problem as well as an effective selection and preparation of the characteristics of the training data during the constitutive modeling of composite materials and the correct application of the ANN.

Keywords: elastic constants of laminated composite materials, artificial neural networks, composite materials, constitutive model of composite materials, training dataset

1. Introduction

Artificial neural networks (ANN) are an efficient artificial intelligence (AI) technique applied in several areas such as bioinformatics [1] for classification, function approximation, and knowledge discovery, as well as for data visualization in medical diagnosis [2].

Various numerical models and experimental techniques have been applied in relation to ANN in the investigation and obtention of the mechanical properties of composite materials, such as Young's Modulus (E), Rigidity Modulus (G), Elastic Limit, and Maximum Tensile Stress [3, 4]. In [5], the different elastics constants of the face-centered cubic austenitic stainless steel are determined. In [6], elastic parameters of an orthotropic material are obtained based on experimental data and using the finite element method (FEM) applied to ANN. The method described in [7] combines the FEM and deep neural networks to obtain constitutive relationships from indirect observations. Acosta et al. [8] use a linear constitutive analytical model proposed in

IntechOpen

[9] for the analysis and obtention of elastic constants of laminated composite materials with metallic layers. The elastic constants of a laminate's component are obtained through an axial load experimental test.

In the constitutive modeling of composite materials, the ANN applications' state of the art, [10] exposes the obstacles that have been encountered due to the difficulty of having a large amount of constitutive experimental training data.

This research presents a method to obtain the elastic constants of one of the components of a MLCM using ANN. The amount of data needed for training was obtained using constitutive models of composite materials proposed by [8].

2. Artificial neural networks (ANN)

ANN are a model inspired by the functioning of the human brain and are made up of connected node set (artificial neurons) that transmit signals to each other from an input stage to generate an output, in order, to improve their learning process by automatically modifying each other. There are several types of neural networks [11, 12] including recurrent neural networks (RNN) and feed-forward neural networks. The latter is an artificial neural network where the connections between the units do not form a cycle and where the information only moves forward.

This research used feed-forward ANN, made up of neurons grouped in layers alongside an input layer, one or more hidden layers, and an output layer. Each network neuron has a weight, a numerical value that modifies the received input. The new modified values are output from the neurons, if the output of any individual neuron is above the specified threshold value the neuron fires and sends data to the next layer of the network, otherwise, the data does not go through. This operation can be appreciated in **Figure 1**.

The H_j neuron has an assigned weight to each of its inputs (Eq. (1)); the assigned weight by H_i to H_j is represented as w_{ij}. The threshold represents the neuron's degree of inhibition, and it is represented as $a_i(t)$. Eq. (2) is calculated with an activation function f(t), which can be linear (Eq. (3)), tangential (Eq. (4)), or hyperbolic tan, (Eq. (5)).

$$H_j = \sum_{n=1}^{i} w_{ij} a_i + B_i \tag{1}$$

$$a_i(t) = f_i(H_i) \tag{2}$$

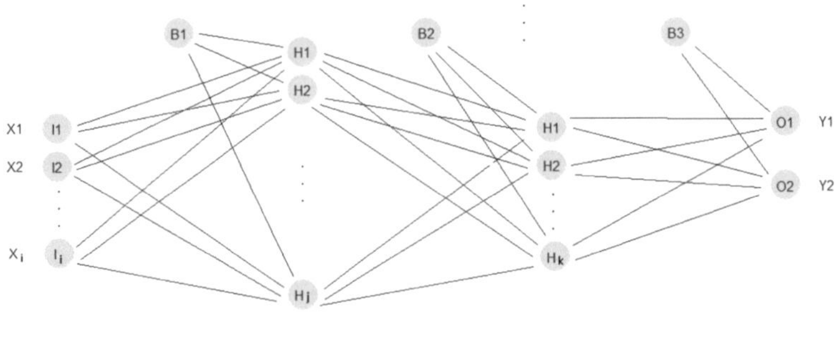

Figure 1.
Example of a feed-forward ANN configuration with i-inputs (X1 … Xi) in the input layer, bias1 and j neurons in hidden layer one (B1, H1, …, Hj), bias2 and k neurons in hidden layer two (B2, H1, …, Hk), and bias3 in the output layer (B3, Y1, Y2) for the two outputs.

$$f(t) = t \tag{3}$$

$$f(t) = \frac{1}{1 + e^{-t}} \tag{4}$$

$$f(t) = \frac{e^{t} - e^{-t}}{e^{t} + e^{-t}} \tag{5}$$

Thus, each ANN neuron, except those in the input layer and the bias neurons, processes all its inputs and provides its own activation as an output.

Once the ANN has been designed, the training process begins to ensure that the w_{ij} given by each neuron is set correctly so that the entire network provides an acceptable output.

During this process, the neural network is capable of storing knowledge from a subset of data containing information on both the inputs and their corresponding outputs, which are known as "desired outputs." The network's obtained outputs are compared with the desired outputs, thus updating the synaptic weights (w_{ij}) so as to reduce the margin of error in the network results. This procedure is repeated until the network reaches a satisfactory performance. One of the used methods to train the ANN is backpropagation [13, 14], where the wij update is done by gradient descent, minimizing the mean squared error (MSE) (Eq. (6)).

$$MSE = \sqrt{\frac{1}{N}\sum_{i=1}^{N}\left(y_{Pred,i} - y_{act,i}\right)^{2}} \tag{6}$$

Overfitting, an ANN flaw [15–17], prevents it from obtaining acceptable outputs from unobserved data, that is, those not used in training. Ying X [18] proposes the following strategies to minimize the effects of overfitting: (1) stop training before finding the optimal MSE; (2) exclude any noise in the training set; and (3) expand the training data.

3. Methodology of determining the elastic constants of a metal laminated composite material using artificial neural networks

Obtaining efficient and consistent results when calculating the elastic constants of a MLCM using an ANN with a constitutive model of composite materials, requires a clear and complete understanding of the analytical model presented in [8] for the efficient preparation of the training data and the correct application of the ANN. The methodology used in this work is as follows:

1. Physical description of the linear analytical model of the axial load of composite laminated materials identifying the role of the implicit variables and parameters present in the model.

2. Definition of the objectives to be solved and identification of the sufficient and necessary parameters that will be used during the training phase. The composition of the composite (position and dimension of the components), the boundary conditions, the geometry, and the dimensions are defined on this stage. The values of the strains are obtained through those parameters and through applying the analytical models.

3. With the application of ANN comes a description of their operation and processes, as well as the characteristics of the training and test data that correspond to the different selected laminated composite materials. The quantitative value of the data in the dataset is delimited depending on the model, the use of the ANN and the final application to obtain one of the MLCM components' elastic constants. The ANN were trained using the R software "neuralnet."

4. The process of applying neural networks to determine elastic constants is carried out using the trained ANN and the data presented in [8].

5. Finally, the analysis of results is carried out by means of the percentages of the relative error (RE) of each of the obtained configurations from the ANN in relation to the data of [8].

4. Analytical model of linear of axial load of composite laminate materials

This study uses a linear analytical model of a composite laminated material made up of layers of metallic material. It is assumed the laminate components are relatively thin, homogeneous, with elastic and linear properties and that the union between them, is perfect.

There is a global uniaxial stress problem, a homogeneous state of strain is considered throughout the laminate as well as in the layers, each point of the laminate presents a state of plane stress.

At the local level, the problem is each layer is biaxial of stresses and the normal stresses have a constant average distribution throughout the thickness of the layers. The state of plane stress generated at the internal points of each of the layers (local analysis) will be referred to as intralaminar state of stress while the stress components of layer i in directions 1 and 2 will be called intralaminar stresses (σ_{xi} and σ_{yi}).

The linear analytical model allows the application of the superposition principle (SP) considering the general problem as a set of individual problems. Therefore, for

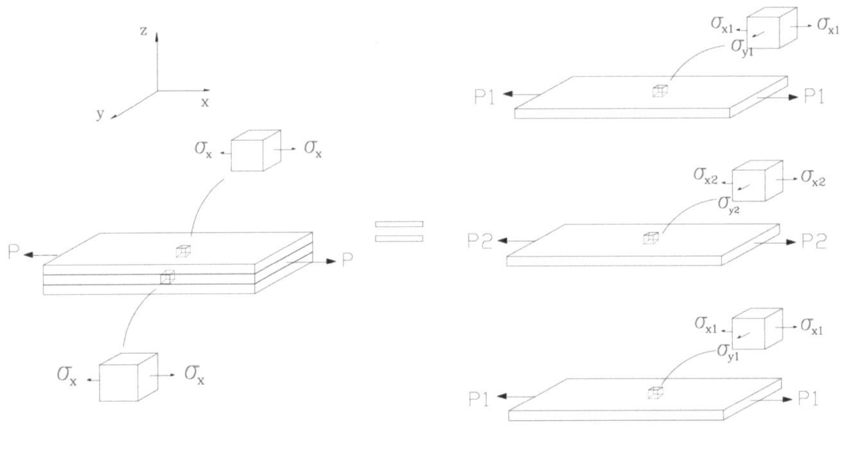

Figure 2.
Representation of the stress state, global, and local models [8].

each load condition, the state of global stresses (average or total) σ_{Gx} and σ_{Gy} are the sum of the states of individual stresses (local) in each layer, see **Figure 2**. The analytical model's global–local equation (Eq. (7)) is as follows:

$$\begin{aligned} \sigma_{Gx} &= n_I\sigma_{xI} + n_{II}\sigma_{xII} + n_{III}\sigma_{xIII} + \ldots + n_i\sigma_{xi} \\ \sigma_{Gy} &= n_I\sigma_{yI} + n_{II}\sigma_{yII} + n_{III}\sigma_{yIII} + \ldots + n_i\sigma_{yi} \end{aligned} \tag{7}$$

σ_{xi} and σ_{yi} represent intralaminar stresses and σ_{Gx} and σ_{Gx} are the global average of the stresses in both x and y directions. n_i represents the volumetric fraction of material layers, $n_I = h_i/h$, where

$$1 = n_I + n_{II} + n_{III} + \ldots + n_i \tag{8}$$

The values of n_i are the volumetric fractions of material with different properties, and h and h_i are both the total thickness and the thickness of the layers or layers groups, respectively.

4.1 Definition of the experimental and illustrative example problem and identification of parameters to consider

The application of the ANN technique requires a data set that helps the network to learn certain patterns related to the analyzed problem. The variables and parameters that will be considered as input and output data during the numerical application of the ANN must be those necessary and sufficient so that the problem is representative. If some key parameters are not considered in the problem, the performed study will be an incomplete and poorly formulated problem, implying a deficient solution.

In a mechanical problem, the state of stress is a function of the position, geometry, boundary conditions, and material. For the discussed problem here, the applied stresses at their σ_{Gx} and σ_{Gy} boundaries were uniformly distributed. Considering the strain state was homogeneous, the state of the plane stresses at a point was independent of the position within each component.

For the analyzed case in [8], which uses a laminated composite material consisting of metallic layers of two different materials (isotropic, homogeneous, and elastic-linear), Eqs. (7) and (8), globally and locally, respectively, are as follows:

$$\begin{aligned} \sigma_{Gx} &= n_1\sigma_{xM1} + n_2\sigma_{xM2} \\ \sigma_{Gy} &= n_1\sigma_{yM1} + n_2\sigma_{yM2} \end{aligned} \tag{9}$$

$$\begin{aligned} \boldsymbol{\sigma}_{\mathbf{xM1}} &= \mathbf{Q}_{\mathbf{11M1}}\boldsymbol{\varepsilon}_{\mathbf{x1}} + \mathbf{Q}_{\mathbf{12M1}}\boldsymbol{\varepsilon}_{\mathbf{y1}} \\ \boldsymbol{\sigma}_{\mathbf{yM1}} &= \mathbf{Q}_{\mathbf{21M1}}\boldsymbol{\varepsilon}_{\mathbf{x1}} + \mathbf{Q}_{\mathbf{22M1}}\boldsymbol{\varepsilon}_{\mathbf{y1}} \\ \boldsymbol{\sigma}_{\mathbf{xM2}} &= \mathbf{Q}_{\mathbf{11M2}}\boldsymbol{\varepsilon}_{\mathbf{x2}} + \mathbf{Q}_{\mathbf{12M2}}\boldsymbol{\varepsilon}_{\mathbf{y2}} \\ \sigma_{\mathrm{yM2}} &= \mathrm{Q}_{\mathrm{21M2}}\varepsilon_{\mathrm{x2}} + \mathrm{Q}_{\mathrm{22M2}}\varepsilon_{\mathrm{y2}} \end{aligned} \tag{10}$$

And considering the engineering constants:

$$\begin{aligned} \boldsymbol{Q_{11M1}} &= \boldsymbol{Q_{22M1}} = \frac{\boldsymbol{E_{M1}}}{\left(\mathbf{1} - \boldsymbol{v_{M1}^2}\right)} \\ \boldsymbol{Q_{12M1}} &= \boldsymbol{Q_{21M1}} = \frac{\boldsymbol{v_{M1}E_{M1}}}{\left(\mathbf{1} - \boldsymbol{v_{M1}^2}\right)} \end{aligned}$$

$$\mathbf{Q}_{11M2} = \mathbf{Q}_{22M2} = \frac{\boldsymbol{E}_{M2}}{\left(\mathbf{1} - \boldsymbol{v}_{M2}^2\right)}$$
$$Q_{12M2} = Q_{21M2} = \frac{v_{M2}E_{M2}}{\left(1 - v_{M2}^2\right)} \tag{11}$$

Eq. (10) can be represented as follows:

$$\sigma_{xM1} = \frac{E_{M1}}{\left(1 - \nu_{M1}^2\right)}\varepsilon_{x1} + \frac{\nu_{M1}E_{M1}}{\left(1 - \nu_{M1}^2\right)}\varepsilon_{y1}$$
$$\boldsymbol{\sigma}_{yM1} = \frac{\boldsymbol{\nu}_{M1}\mathbf{E}_{M1}}{\left(\mathbf{1} - \boldsymbol{\nu}_{M1}^2\right)}\boldsymbol{\varepsilon}_{x1} + \frac{\mathbf{E}_{M1}}{\left(\mathbf{1} - \boldsymbol{\nu}_{M1}^2\right)}\boldsymbol{\varepsilon}_{y1}$$
$$\boldsymbol{\sigma}_{xM2} = \frac{\mathbf{E}_{M2}}{\left(\mathbf{1} - \boldsymbol{\nu}_{M2}^2\right)}\boldsymbol{\varepsilon}_{x2} + \frac{\boldsymbol{\nu}_{M2}\mathbf{E}_{M2}}{\left(\mathbf{1} - \boldsymbol{\nu}_{M2}^2\right)}\boldsymbol{\varepsilon}_{y2}$$
$$\sigma_{yM2} = \frac{\nu_{M2}E_{M2}}{\left(1 - \nu_{M2}^2\right)}\varepsilon_{x2} + \frac{E_{M2}}{\left(1 - \nu_{M2}^2\right)}\varepsilon_{y2} \tag{12}$$

Here, Q_{11M1}, Q_{12M1}, Q_{22M1}, Q_{11M2}, Q_{12M2}, and Q_{22M2} represent the material's stiffness constants. The engineering constants for each layer were Young's moduli (E_{M1} and E_{M2}) and Poisson's ratios (ν_{M1} and ν_{M2}). The deformation states were defined for each layer through their longitudinal strains: ε_{x1}, ε_{y1} and ε_{x2}, ε_{y2}.

5. Neural network training process

5.1 ANN arguments

As mentioned in Sections 2 and 3, the used software to train the ANNs was R [19] and the used library was "neuralnet" [20], the used parameters are shown in **Table 1**.

Formula	Description of the model
data	Dataset of variables specified in formula
hidden	Number of hidden layers and number of neurons
stepmax	Maximum steps for the training
threshold	Value for the error function as stopping criteria
rep	Number of repetitions for the training
startweights	Starting values for the weights
learningrate	Lowest and highest limit for the learning rate
algorithm	Name of type to calculate the ANN
err.fct	Function that is used for error
act.fct	Name of the activation function
linear.output	Boolean value for output layer
constant.weights	The weights that are exclude from training

Table 1.
Arguments for the neuralnet function.

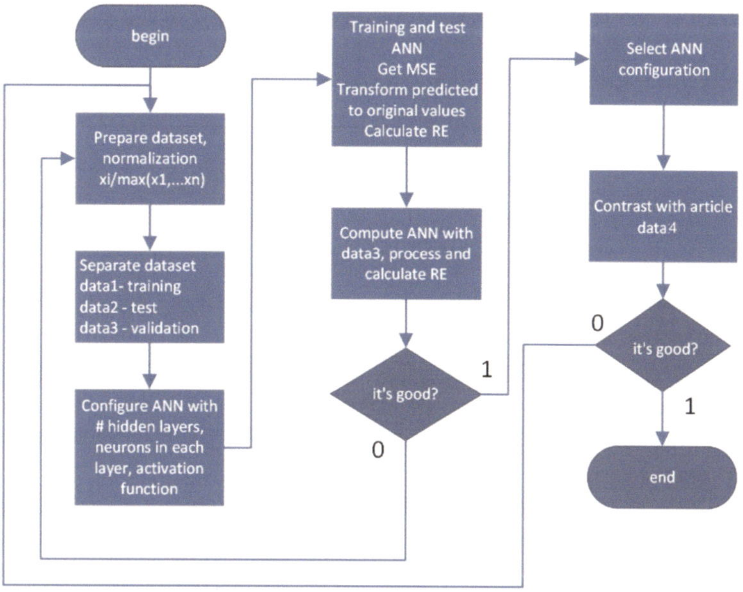

Figure 3.
Procedure diagram for the ANN training process.

The used learning algorithm was resilient backpropagation [21, 22], which modifies the updated values for each weight, w_{ij} according to the sign sequence behavior of the partial derivative equations in each dimension of the weight space, this reduces the number of steps compared to the original gradient descent backpropagation procedure.

The procedure to obtain a good ANN begins with the generation of the dataset through a normalization process that allows scaling the data values to improve learning. The process utilized scaling over the maximum value of the inputs as seen in Eq. (13).

$$x_i = \frac{x_i}{max\,(x_1,\ \ldots,x_n)} \forall i \in \{1,\ \ldots,n\} \tag{13}$$

The best and final dataset built for this study consisted of 253 pieces of data, 76% of which were used for training (data1), 14% for testing (data2), and the remaining 10% for model validation (data3) (192, 36, 25). A final dataset (data4) was built for the ANN application as indicated in Section 7.3 with results as published in [8].

The variants in the arguments for ANN generation in this study were 2 or 3 hidden layers, with either the hyperbolic tangent (tanh) activation functions (Eq. (5)) or the logistic function (Eq. (4)) The number of neurons in each hidden layer was chosen to obtain the lowest RE in both the training dataset and the test dataset. All this is depicted in **Figure 3**.

It should be noted that some variations in the structural presentation of the inputs and outputs were made for the elaboration of the dataset, this was necessary since high MSE values were obtained during the ANN training.

5.2 Generation of training data from the analytical model for the ANN

As seen in Eqs. (10) and (12), the necessary and sufficient variables that define the plane stress models, based on the stiffness constants and the engineering constants, are:

a. the material's volumetric fraction of the components in the laminate (n_i).

b. the stress components of the global stress state (σ_{Gx} and σ_{Gy}).

c. the local stress state in each component (σ_{xi} and σ_{yi}).

d. the elastic constants of the known components (E_{M1}, E_{M2}, ν_{M1}, and ν_{M2} or Q_{11M1}, Q_{12M1}, Q_{11M2}, and Q_{12M2}).

e. the strains state equal for all points of the laminate: ε_{x1}, ε_{y1}, ε_{x2}, and ε_{y2}.

As seen in Eqs. (10) and (12), the necessary and sufficient variables that define the state of plane stress models based on the stiffness constants and engineering constants are the material concentration of the components in the laminate (n_i); the stress components of the global state of stress σ_{Gx} and σ_{Gy}; the local states of stress in each component (σ_{xi} and σ_{yi}); and the elastic constants of the known components (E_{M1}, E_{M2}, ν_{M1}, and ν_{M2} or Q_{11M1}, Q_{12M1}, Q_{21M1}, and Q_{22M1}).

When the ANN objective is directly to determine the engineering elastic constants of one of the components of the laminate, the input parameters are Eqs. (9) and (12): n_1, n_2, σ_{Gx}, σ_{Gy}, ε_{x1}, ε_{y1}, ε_{x2}, ε_{y2}, E_{M1}, ν_{M1} and outputs E_{M2}, ν_{M2}, the EvANN was constructed. And a QANN was constructed for Eqs. (9) and (10): with input parameters n_1, n_2, σ_{Gx}, σ_{Gy}, ε_{x1}, ε_{y1}, ε_{x2}, ε_{y2}, Q_{11M1}, and Q_{12M1}. and outputs Q_{11M2} and Q_{12M2}.

5.3 Specification of quantitative ranges of input data

As described in the methodology, the input data must establish:

a. The parameters that define the problem, identifying inputs and outputs.

b. The quantitative ranges in the boundary conditions, global stresses (σ_{Gx} and σ_{Gy}).

c. The quantitative ranges in the volumetric fractions of each of the components in the laminate (n_1 and n_2).

d. The elastics constants (E and ν) of the MLCM component materials.

e. The global strains (ε_{xM1}, ε_{yM1} and ε_{xM2}, ε_{yM2}) for a simple tension problem. These dependent and necessary data for training were obtained from the analytical model. Eqs. (10) and (12) were solved using MAPLE 2018 [23], for strains, see Eqs. (14) and (15).

$$\varepsilon_x = \frac{\sigma_{Gx}(Q_{11M1}n_1 + Q_{11M2}n_2)}{Q_{11M1}^2 n_1^2 + 2Q_{11M1}Q_{11M2}n_1n_2 + Q_{11M2}^2 n_2^2 - Q_{12M1}^2 n_1^2 - 2Q_{12M1}Q_{12M2}n_1n_2 - Q_{12M2}^2 n_2^2}$$

$$\varepsilon_y = -\frac{\sigma_{Gx}(Q_{12M1}n_1 + Q_{12M2}n_2)}{Q_{11M1}^2 n_1^2 + 2Q_{11M1}Q_{11M2}n_1n_2 + Q_{11M2}^2 n_2^2 - Q_{12M1}^2 n_1^2 - 2Q_{12M1}Q_{12M2}n_1n_2 - Q_{12M2}^2 n_2^2} \tag{14}$$

and

$$\begin{aligned} \boldsymbol{\varepsilon_x} &= \frac{\boldsymbol{\sigma_{Gx}\left(E_{M1}n_1\nu_{M2}^2 + E_{M2}n_2\nu_{M1}^2 - E_{M1}n_1 - E_{M2}n_2\right)}}{\boldsymbol{E_{M1}^2n_1^2\nu_{M2}^2 + 2E_{M1}E_{M2}n_1n_2\nu_{M1}\nu_{M2} + E_{M2}^2n_2^2\nu_{M1}^2 - E_{M1}^2n_1^2 - 2E_{M1}E_{M2}n_1n_2 - E_{M2}^2n_2^2}} \\ \varepsilon_y &= -\frac{\sigma_{Gx}\left(E_{M1}n_1\nu_{M2}^2 + E_{M2}n_2\nu_{M1}^2\nu_{M2} - E_{M1}n_1\nu_{M1} - E_{M2}n_2\nu_{M2}\right)}{E_{M1}^2n_1^2\nu_{M2}^2 + 2E_{M1}E_{M2}n_1n_2\nu_{M1}\nu_{M1} + E_{M2}^2n_2^2\nu_{M1}^2 - E_{M1}^2n_1^2 - 2E_{M1}E_{M2}n_1n_2 - E_{M2}^2n_2^2} \end{aligned} \tag{15}$$

The maximum and minimum quantitative value of the boundary conditions in the training data was established using the values found in [8] as a reference. Between 1 and 22 MPa for the global input stress. The components concentrations in the MLCM were bounded for values between 0 and 1 for 2, 3, 4, 5, and 6 layers of two metallic components assumed to have the same thickness. **Tables B1–B4** in Appendix B show various scenarios evaluated during the study.

The considered scenarios were:

a. Different MLCM configurations with different concentrations and different components.

b. Different global stress values (σ_{Gx} and σ_{Gy}).

c. Two ANN targets for each configuration, one to obtain the elastic constants of one of two MLCM components M1, and another to obtain the constants of M2.

The obtained training data from the model were adjusted so that there was not much difference in the order of the values, the stress was given in MPa, the Q's and E's in GPa and the strains in $\mu\varepsilon$.

An EvANN model to determine engineering constants and another to determine stiffness coefficients, QANN model, were presented to contextualize the effect that occurs when an ANN model is trained from simple knowledge or general knowledge, their implications can be seen in Eqs. (14) and (15).

The nomenclature used in the analytical model and the ANN network formulas is shown in Appendix A **Table A1**.

5.4 EvANN

As mentioned above, this ANN was trained using the engineering constants and the R software. The settings for the "neuralnet" function are given in **Table 2**, where the output variables are the second material constants.

Starting from the first dataset training was carried out obtaining a MSE of 1.186e +09 and 4.391 for unnormalized and normalized data. Because of this, the dataset was extended considering a larger number of MLCM configurations with variations in n concentrations and global stress ranges, as well as, for the same mechanical problem, was done an inverted request in the elastic constants of component M1, for one case, and M2 for another.

Table 3 shows the configured EvANNs and specifies the activation function, the number of hidden layers, the number of neurons in each layer, the MSE, the threshold reached, and the number of steps performed. The dataset used can be found in Appendix B.

Formula	EM2 + vM2 ~ SGX + SGY + CON1 + CON2 + EX + EY + EM1 + vM1
data	dataset (192 training, 36 test, 25 validation)
hidden	c(a,b) or c(a,b,c) where a,b,c are the number of neurons
stepmax	1.00E+07
threshold	0.01
rep	1
startweights	NULL
learningrate	0.0001
algorithm	rprop (resilient backpropagation)
err.fct	sse (sum of squared errors)
act.fct	tanh (tangent hyperbolicus) or logistic (logistic function)
linear.output	TRUE
const.weights	TRUE

Table 2.
"Neuralnet" Argument functions for training EvANN.

ID	Activation function	Hidden layers	Neurons per layer	MSE ANN	Reached threshold	Steps
1	tanh	2	12,4	0.04742882	0.00801973	5612
2	logistic	2	12,4	0.01112887	0.009550541	15,577
3	tanh	2	12,6	0.02183281	0.009768165	30,691
4	logistic	2	12,6	0.03428463	0.009515263	7269
5	tanh	2	12,8	0.05968034	0.009165002	29,721
6	logistic	2	12,8	0.02881464	0.008727336	4276
7	tanh	2	14,8	0.04617885	0.008214281	37,203
8	logistic	2	14,8	0.03225161	0.009175114	4013
9	tanh	3	14,6,4	0.01818992	0.009329059	3394
10	tanh	3	12,8,4	0.01749427	0.009955628	5562
11	tanh	2	18,8	0.04391595	0.00908179	12,742

Table 3.
EvANN Configurations with different activation function, hidden layer, and number of neurons.

The third EvANN configuration and its graph is shown in **Figure 4** along with the relative error percentages in **Figures 5** and **6**, the RE was calculated with Eq. (16).

$$RE = \left|\frac{\text{Real Value} - \text{ANN Value}}{\text{Real Value}}\right| * 100 \tag{16}$$

A second ANN was generated for the same problem now defined in terms of Q11 and Q12.

DOI: http://dx.doi.org/10.5772/intechopen.108601

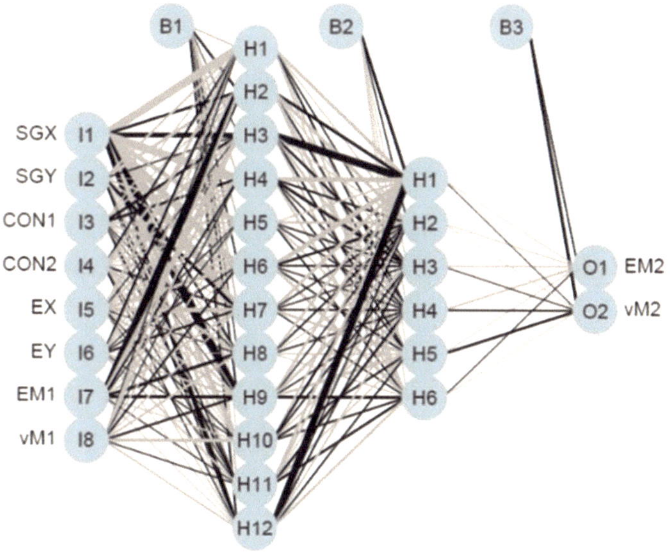

Figure 4.
EvANN topology image. Input layers and (EM2) and (vM2) in output layers.

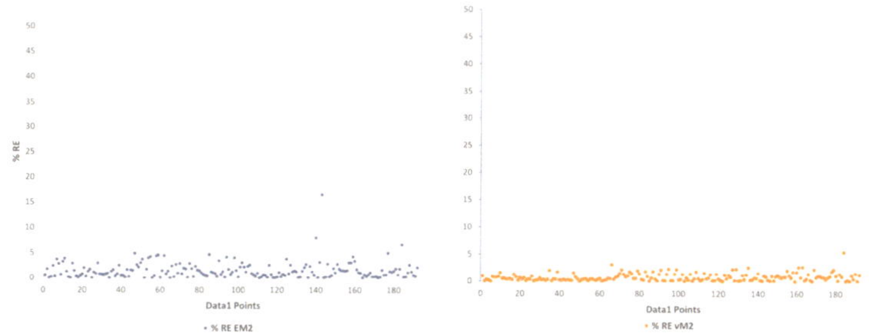

Figure 5.
% RE training EvANN Configuration 3.

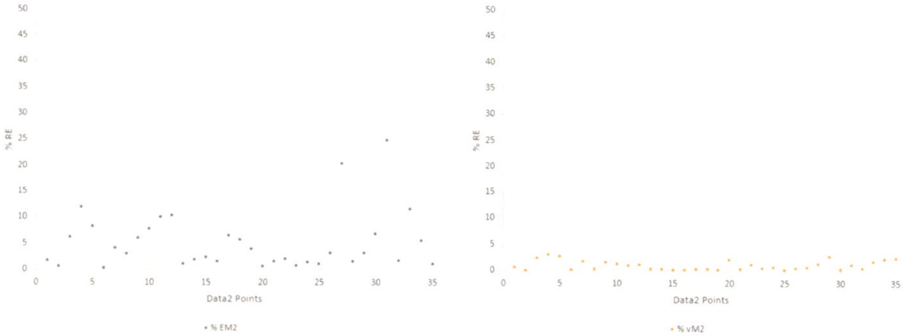

Figure 6.
%RE, test dataset ANN Configuration 3.

5.5 QANN

A second model with stiffness coefficients was now developed with the results being the coefficients of the second material, Q11M2 and Q12M2, **Figure 7** and **Table 4**. The settings for the "neuralnet" function are given in **Table 4**.

The generated configurations with their respective achieved values are shown in **Table 5**.

The second QANN configuration and its graph is showed in **Figure 7** with the relative error percentages, **Figures 8** and **9**, RE which computes with Eq. (16).

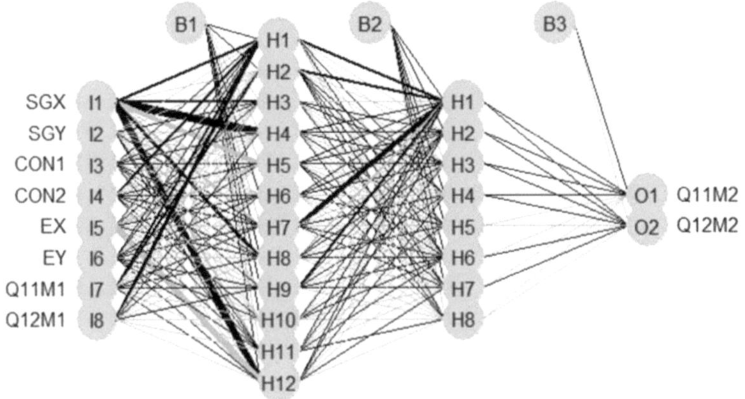

Figure 7.
QANN topology image. Input layers and Q11 Material 2 (Q11M2); Q12 Material 2 (Q12M2) in output layer configuration 2.

Formula	**Q11M2 + Q12M2 ~ SGX + SGY + CON1 + CON2 + EX + EY + Q11M1 + Q12M1**
data	dataset (192 training, 36 test, 25 validation)
hidden	c(a,b) or c(a,b,c) where a,b,c are the number of neurons
stepmax	1.00E+07
threshold	0.01
rep	1
startweights	NULL
learningrate	0.0001
algorithm	rprop (resilient backpropagation)
err.fct	sse (sum of squared errors)
act.fct	tanh (tangent hyperbolics) or logistic (logistic function)
linear.output	TRUE
constant.weight	TRUE

Table 4.
"Neuralnet" argument functions for training QANN.

ID	Activation function	Hidden layers	Neurons	MSE ANN	Reached threshold	Steps
1	tanh	2	12,6	3.09E-02	9.93E-03	4.49E+04
2	tanh	2	12,8	2.03E-01	8.57E-03	5.50E+04
3	logistic	2	12,8	5.02E-02	9.41E-03	2.15E+04
4	tanh	2	16,6	2.01E-01	7.51E-03	2.08E+04
5	logistic	3	16,6	5.02E-02	9.90E-03	2.67E+04
6	tanh	3	16,6,4	3.43E-03	7.57E-03	1.17E+04
7	logistic	3	16,6,4	5.68E-03	9.70E-03	5.52E+03

Table 5.
QANN Configurations with different activation function, hidden layer, and number of neurons.

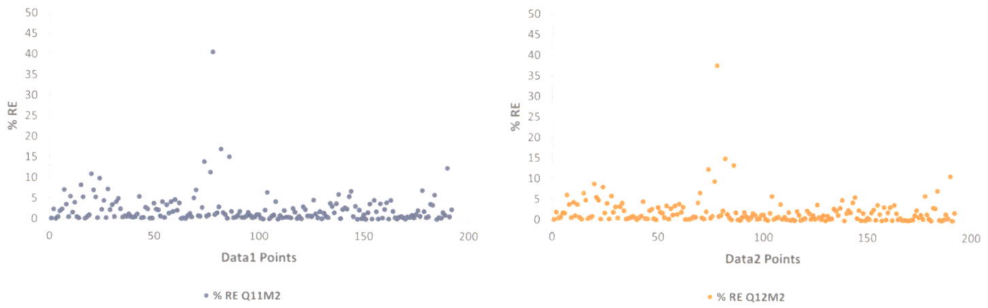

Figure 8.
% RE training QANN Configuration 2.

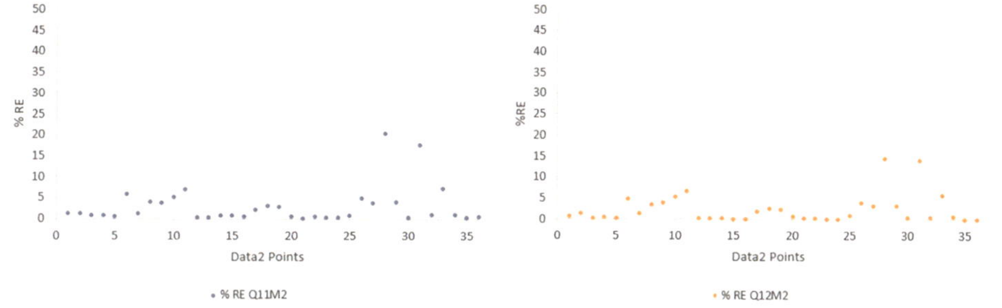

Figure 9.
% RE, test dataset QANN Configuration 2.

6. Neural network validation process

Once an ANN has been trained and tested, it is evaluated by applying it to an equivalent problem but with different structural values from those used in the training. The results for the different scenarios are in **Table 6** and **Figures 10** and **11**.

Configurations 3 and 2 were selected for the evaluation of both, EvANN and QANN, based on their performance in the test dataset.

Table 6 and **Figures 10** and **11** depicts the configuration and attributes selected for each ANN.

The RE for each ANN is shown in the plots in **Figures 10** and **11**.

	Activation function	Hidden layers	Neurons	MSE ANN	Reached threshold	Steps
EvANN	tanh	2	12,6	2.18E-02	9.83E-03	3.07E+04
QANN	tanh	2	12,8	2.03E-01	8.57E-03	5.50E+04

Table 6.
Configurations selected for EvANN and QANN.

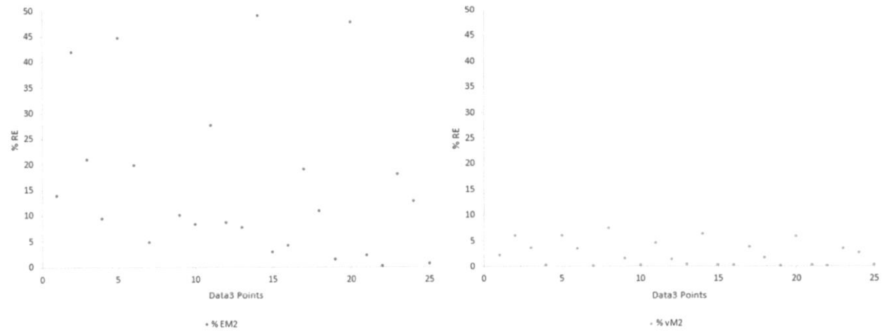

Figure 10.
% RE validation dataset EvANN Model.

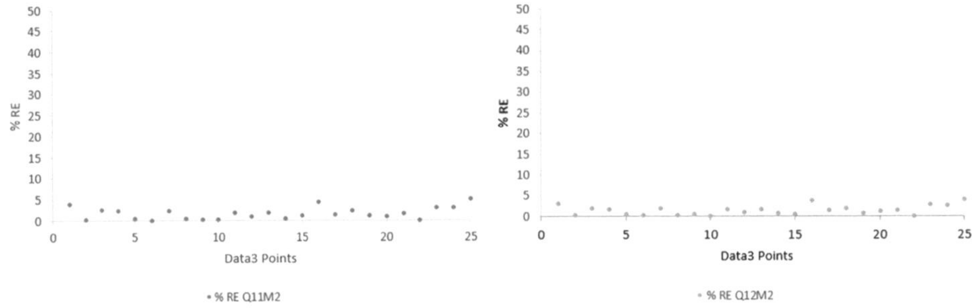

Figure 11.
% RE QANN validation dataset.

As can be seen in the tables, the maximum RE obtained are up to 58.5% for the EM2 output and up to 7.5% for vM2.

When looking at QANN, the tables show the maximum RE obtained was up to 4.48% for Q11M2 production and up to 3.71% for Q12M2.

7. Results of application

In this section, the contrasting results of EvANN and QANN RE for data3 and compute with the results published in the article [8], data4 are presented .

7.1 Validation process (Data3)

The training data was expanded and a configuration that is not close to the optimal MSE value was selected to avoid overfitting. For the EvANN, this occurred with

	Young's Modulus E			Poisson's Ratio v		
	Constants model of three material	**Constants evaluated**	**% RE**	**Constants model**	**Constants evaluated**	**% RE**
EvANN	67	65.45	2.31	0.33	0.3293	0.22
QANN		67.08	0.12		0.3298	0.05
EvANN	100	61.76	38.24	0.31	0.3266	5.37
QANN		101.02	1.02		0.3102	0.05
EvANN	200	178.04	10.98	0.29	0.2952	1.80
QANN		202.25	1.12		0.2902	0.06

Table 7.
Comparison of results for EvANN and QANN for three materials in dataset 3 for the engineering constants.

configuration 3, which has a higher MSE than configurations 2, 9, and 10, as shown in **Table 3**. For the QANN, configuration 2 was selected, which has two hidden layers with 12 and 8 neurons, respectively, with the tanh activation function (hyperbolic tangential), as shown in **Table 5**. The selection criterion was to obtain the lowest average RE in the two output variables.

The RE of EvANN and QANN outputs is shown in **Table 7**, where the QANN outputs were converted to engineering constants (Eqs. (17) and (18)).

$$E_{M2} = \frac{Q_{11M2}^2 - Q_{12M2}^2}{Q_{11M2}} \quad (17)$$

$$\nu_{M2} = \frac{Q_{12M2}}{Q_{11M2}} \quad (18)$$

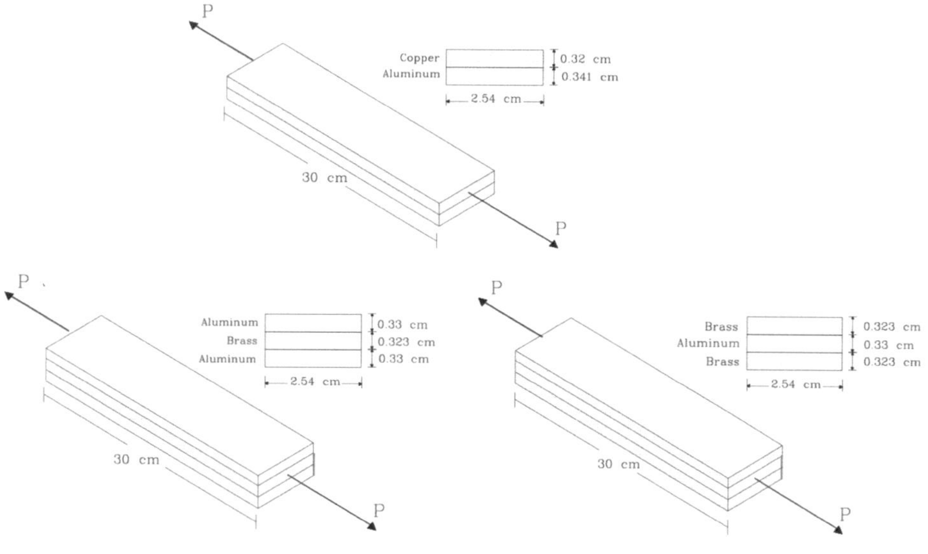

Figure 12.
MLCM test tubes used in the final ANN application.

MCLM	Aluminum volumetric fractions, n	Brass volumetric fractions, n	Aluminum volumetric fractions, n	Copper volumetric fractions, n
Aluminum-brass-aluminum	0.671	0.329		
Brass-aluminum-brass	0.338	0.662		
Aluminum-copper			0.516	0.484

Table 8.
Volumetric fractions of materials in the MLCMs.

Material	Young's Modulus (E), Gpa	Poisson's Ratio (υ)
Aluminum	67	0.345
Brass	101	0.313
Copper	109	0.33

Table 9.
Elastic constants of the MLCM components.

7.2 Real case data (Data4)

The different MLCMs used in this stage (**Figure 12**) are (1) Aluminum-Brass-Aluminum (A-B-A), (2) Brass-Aluminum-Brass (B-A-B) and Copper-Aluminum (C-A) (**Figure 3**). The properties and volumetric fractions of the materials in the MLCM are given in **Tables 8** and **9**.

7.3 Real case data compute in QANN and EvANN

Continuing with the real case, the contrasted RE of EvANN and QANN results are presented as shown in **Figure 13**. The QANN model shows a better performance.

Table 10 shows the averages of the RE obtained for each of the outputs when checking the application of the model obtained using QANN for the results in [8].

Configuration two, which has two hidden layers with 12 and 8 neurons in relation to the tanh (hyperbolic tangential) activation function, was selected. The selection

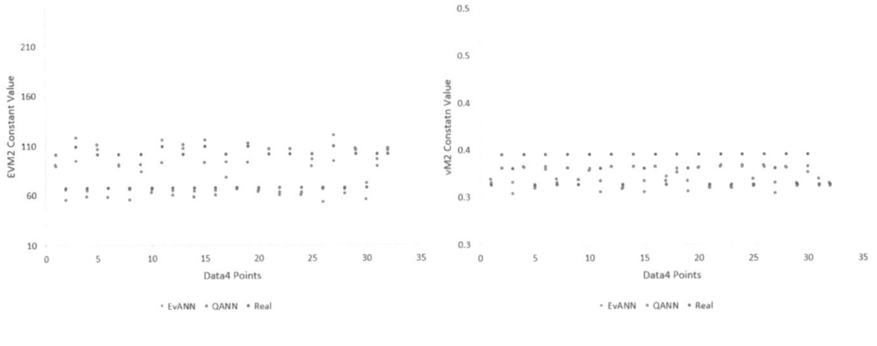

Figure 13.
Comparison of EvANN and QANN results based on engineering constants, EVM2, vM2.

ID	Activation function	Num layers	Neurons	Q11M2				Q12M2			
				RE media	Standard deviation	Min	Max	RE media	Standard deviation	Min	Max
1	tanh	2	12,6	8.72	9.86	0.72	50.00	37.20	80.76	0.84	301.50
2	tanh	2	12,8	5.46	4.37	0.34	21.87	4.63	3.89	0.18	18.82
3	logistic	2	12,8	12.17	8.15	0.91	49.29	11.45	6.07	0.88	33.17
4	tanh	2	16,6	5.33	6.68	0.37	36.97	5.30	5.63	0.72	31.28
5	logistic	3	16,6	7.28	10.18	0.22	55.73	6.35	7.01	0.13	36.56
6	tanh	3	16,6,4	5.64	12.40	0.06	71.24	5.40	9.57	0.10	51.09
7	logistic	3	16,6,4	6.63	14.73	0.23	84.72	6.31	10.71	0.37	57.61

Table 10.
Means and standard deviation of RE for different configurations of the QANN application with the article results.

ANN, Applied	Average RE %	
	Constants EM2	Constants vM2
EvANN	12.54	3.15
QANN	6.184	3.58

Table 11.
Average percentages RE for the stiffness and engineering constants.

criterion was to obtain the smallest average percentage of error in the two output variables, Q11M2 and Q12M2 as in Section 7.1, which obtained smaller RE for data3 using the QANN.

Table 11 shows the results and the contrast in RE average percentage of all the considered outputs from the final results, obtained for the ANN, for the stiffness constants (QANN) and for the engineering constants (EvANN).

The values of the engineering constants as a function of Q were determined from the identity equations (Eqs. (17) and (18)), and the obtained average for each of the specimens in **Table 12**. The table also shows that the results for each configuration line should be very approximate since these were acquired from linear multiplication, but have variations.

Finally, the QANN configuration that shows the best results during validation was applied to the MLCMs analyzed in [8]. The value of the final average constants is presented in **Table 13**.

8. Discussion

The different conditions described in Section 5, were evaluated as a result of the used study to obtain a trained and efficient ANN, the most important of which were the following:

1. The values of all data used in the training were selected and restricted in such a way that their values, corresponding to the MLCM of [8] (Aluminum-brass-

Configuration line (Brass-Aluminum)	**Young's Modulus (E)**						**Poisson's Relation (υ)**					
	EM2 Brass (ANN) (Gpa)	**Expt Brass EM2 (Gpa)**	**RE %**	**EM2 Aluminum (ANN) (Gpa)**	**Expt Aluminum EM2 (Gpa)**	**RE %**	**vM2 Brass (ANN)**	**Expt Brass vM2**	**RE %**	**vM2 Aluminum (ANN)**	**vM2 Aluminum Expt**	**RE %**
1	89.77	101	12.5	64.74	67	3.5	0.315	0.313	0.6	0.332	0.345	4.0
2	83.72	101	20.6	64.64	67	3.6	0.318	0.313	1.6	0.332	0.345	3.9
3	77.47	101	30.4	65.37	67	2.5	0.322	0.313	2.8	0.331	0.345	4.1
4	96.21	101	5.0	65.70	67	2.0	0.312	0.313	0.4	0.331	0.345	4.3
		Average	17.1		Average	2.9		Average	1.4		Average	4.1
Configuration line (Aluminum-Brass-Aluminum)	Young's Modulus (E)						Poisson Relation (υ)					
	EM2 Brass (ANN) (Gpa)	Expt Brass EM2 (Gpa)	RE %	EM2 Aluminum (ANN) (Gpa)	Expt Aluminum EM2 (Gpa)	RE %	vM2 Brass (ANN)	Expt Brass vM2	RE %	vM2 Aluminum (ANN)	Expt Aluminum vM2	RE %
1	106.61	101	5.3	65.64	67	2.1	0.311	0.313	0.6	0.331	0.345	4.2
2	106.89	101	5.5	65.86	67	1.7	0.310	0.313	0.9	0.331	0.345	4.4
3	106.44	101	5.1	65.85	67	1.8	0.310	0.313	1.1	0.330	0.345	4.5
4	105.39	101	4.2	65.90	67	1.7	0.312	0.313	0.3	0.332	0.345	4.0
		Average	5.0		Average	1.8		Average	0.7		Average	4.3
Configuration line (Aluminum-Copper)	Young's Modulus (E)						Poisson Relation (υ)					
	EM2 Aluminum (ANN) (Gpa)	Expt Aluminum EM2 (Gpa)	RE %	EM2 Copper (ANN) (Gpa)	Expt Copper EM2 (Gpa)	RE %	vM2 Aluminum (ANN)	Expt Aluminum vM2	RE %	vM2 Copper (ANN)	Expt Copper vM2	RE %
1	67.25	67	0.4	118.50	109	8.0	0.329	0.313	5.0	0.304	0.330	8.4
2	64.25	67	4.3	115.88	109	5.9	0.332	0.313	5.8	0.305	0.330	8.1
3	62.47	67	7.2	112.54	109	3.1	0.334	0.313	6.3	0.306	0.330	7.8
4	71.67	67	6.5	120.21	109	9.3	0.326	0.313	3.9	0.304	0.330	8.6
		Avg	4.6		Avg	6.6		Avg	5.3		Avg	8.2

Table 12.
Values of the constants of the analytical model [8] and each output for each configuration line.

	Young's Modulus (E)					Poisson's Ratio (υ)				
	Expt constant	Article constant	RE %	Evaluated constant	RE %	Expt constant	Article constant	RE %	Evaluated constant	RE %
Aluminum-Brass-Aluminum										
Aluminum	67	72	7.5	65.11	2.8	0.345	0.34	1.4	0.34	1.4
Brass	101	97.6	3.4	86.8	14.1	0.313	0.318	1.6	0.3148	0.6
Brass-Aluminum-Brass										
Aluminum	67	72	7.5	65.81	1.8	0.345	0.34	1.4	0.33	4.3
Brass	101	97.6	3.4	106.33	5.3	0.313	0.318	1.6	0.311	0.6
Aluminum-Copper										
Aluminum	67	64.4	3.9	66.41	0.9	0.345	0.33	4.3	0.34	1.4
Copper	109	106.2	2.6	116.8	7.2	0.33	0.32	3	0.3148	4.6

Table 13.
Average final elastic constants obtained with QANN.

aluminum (A-B-A), Brass-Aluminum-brass (B-A-B), and Copper-aluminum (C-A) were located at a mean value. An observed case that showed the need for this step was in the training validation, the stress values (σ_{Gx} and σ_{Gy}) in the input were out of range compared to those used in the training, the results had larger differences between 20 and 80%, from when these were close to the mean.

2. The selected materials for training were randomly taken from the literature as only values approximating those needed for the final ANN application were needed (see Appendix B **Table B1**).

3. The homogeneous unit strains were obtained from the model, Eqs. (10) and (12), the MLCM configuration, the elastic constants of its components and the different values of global stresses applied on their boundaries. Thus, the training tables and the ANN validation, Appendix B **Table B2**, were generated.

4. The data for the same MLCM and the same boundary conditions were duplicated for training purposes by inverting the requested outputs: in one case, the objective elastic properties of one of the MLCM components and, in another, those of the other component. This is all depicted in Appendix B **Table B1**. It is important to mention that this activity was relevant because it improved the ANN results by 4.36% in the MSE, where the average of the EvANN configurations is 0.03283643, **Table 4**.

5. Regarding the real case, only one data was available for each MLCM, but since it was a linear mechanical problem, the values were multiplied in boundary conditions and four input configuration lines were obtained for the same piece of data. The presented results in Appendix B, **Table B4** showed the importance of this step, because although these were performed only four times, it was observed that for the same case of the MLCM (A-L-A) the results that ANN had,

were different with variations of up to 30%, when these should be the same. However, uncertainty risks are avoided by averaging the obtained values for each output as shown in **Table 13**.

6. A further important point that showed an overview of simplicity in the training setup was found when evaluating two cases: one in the training phase and the other in the output data request. The first one required the engineering constants while the latter required the stiffness constants. In the first case average RE of 12.54% E and 3.15 for υ were obtained; in the second case, the RE for E and υ were 6.18% and 3.57%, respectively. From the above and observing (Eqs. (14) and (15)), it is assumed that the analytical model in terms of Q's is simpler than the model in terms of engineering constants.

9. Conclusions

This chapter, using ANN, establishes a method to determine the engineering constants of metallic laminated composite material layers, shows the importance of adequately defining the problem to be solved, analyzing concepts, establishing scopes and constraints, and selecting sufficient and necessary training parameters, based on the obtained results. The importance of the following was identified by evaluating several scenarios to generate the ANN dataset: (a) the qualitative ranges of the parameters in the input data; recommending that the values of the application data should be in the mean of the training data, (b) variations in the structure of the dataset (different outputs for the same MLCM problem), and (c) simplicity in the dataset; the ANN showed better results when stiffness constants were requested in the output data; the analytical solution, is simpler in terms of stiffness constants than in terms of engineering constants.

Several configurations with different activation functions, number of layers, and number of neurons per layer were tested in the study, finding better results for this problem with a medium MSE when compared with the lowest MSE trained. This action may be due to the fact that there is no overfitting.

Based on this research, it is recommended to use the analytical model applied here to generate an ANN dataset for the study of the constitutive modeling of composite materials in plane stress problems.

Nomenclature

ANN	Artificial neural network
MLCM	Metal laminated composite material
MCL	Metallic composite
E	Young's Modulus
G	Rigidity Modulus
MSE	Mean squared error
EvANN	ANN that directly determine the engineering elastic constants of one of the components of the laminate
QANN	ANN to determine the stiffness coefficients of one of the components of the laminate

DOI: http://dx.doi.org/10.5772/intechopen.108601

Appendix A:

Parameter	Name in Model	Name in ANN formula
Global stress in x (MPa)	σ_{Gx}	SGX
Global stress in y (MPa)	σ_{Gy}	SGX
Material concentration coefficient 1	n_1	CON1
Material concentration coefficient 2	n_2	CON2
Strain in x	ε_x	EX
Strain in y	ε_y	EY
Young's Modulus, Material 1 (GPa)	E_{M1}	EM1
Young's Modulus, Material 2 (GPa)	E_{M2}	EM2
Poisson's Relation, Material 1	ν_{M1}	vM1
Poisson's Relation, Material 2	ν_{M2}	vM2
Stiffness constants, material 1	Q_{11M1}, Q_{12M1}	Q11M1, Q12M1
Stiffness constants, material 1	Q_{11M2}, Q_{12M2}	Q11M2, Q12M2

Table A1.
Nomenclature used for the analytical model and ANN formula.

Appendix B:

Inputs						Expected outputs			
Global Stress in x (σx) Pa	Global Stress in y (σy) Pa	Volumetric fraction, material 1 (n1)	Volumetric fraction, material 2 (n2)	Strain in x (Ɛx)	Strain in y (Ɛy)	Q11 Material 1 (Q11M1) GPa	Q12 Material 1 (Q12M1) GPa	Q11 Material 2 (Q11M2) GPa	Q12 Material 2 (Q12M2) GPa
1	0	0.666	0.333	0.0090	−0.0028	75.19	24.81	218.36	63.33
1	0	0.333	0.666	0.0090	−0.0028	218.36	63.33	75.19	24.81
2	0	0.666	0.333	0.0120	−0.0035	218.36	63.33	110.63	34.30
2	0	0.333	0.666	0.0120	−0.0035	110.63	34.30	218.36	63.33
2.5	0	0.666	0.333	0.0225	−0.0069	75.19	24.81	218.36	63.33
2.5	0	0.333	0.666	0.0225	−0.0069	218.36	63.33	75.19	24.81
3	0	0.666	0.333	0.0337	−0.0106	110.63	34.30	75.19	24.81
3	0	0.333	0.666	0.0337	−0.0106	75.19	24.81	110.63	34.30
4	0	0.666	0.333	0.0359	−0.0110	75.19	24.81	218.36	63.33
4	0	0.333	0.666	0.0359	−0.0110	218.36	63.33	75.19	24.81
4.5	0	0.666	0.333	0.0506	−0.0159	110.63	34.30	75.19	24.81
4.5	0	0.333	0.666	0.0506	−0.0159	75.19	24.81	110.63	34.30
5	0	0.666	0.333	0.0300	−0.0088	218.36	63.33	110.63	34.30
5	0	0.333	0.666	0.0300	−0.0088	110.63	34.30	218.36	63.33

Table B1.
Part of the data used in the training, (data1).

Global Stress in x (σx) Pa	Global Stress in y (σy) Pa	Volumetric fraction, material 1 (n1)	Volumetric fraction, material 2 (n2)	Strain in x (Ɛx)	Strain in y (Ɛy)	Q11 Material 1 (Q11M1) GPa	Q12 Material 1 (Q12M1) GPa	Q11 Material 2 (Q11M2) GPa	Q12 Material 2 (Q12M2) GPa
3.5	0	0.666	0.333	0.0210	−0.0062	218.36	63.33	110.63	34.30
5.5	0	0.666	0.333	0.0494	−0.0151	75.19	24.81	218.36	63.33
8	0	0.666	0.333	0.0480	−0.0141	218.36	63.33	110.63	34.30
9.5	0	0.666	0.333	0.0571	−0.0168	218.36	63.33	110.63	34.30
11	0	0.666	0.333	0.0661	−0.0194	218.36	63.33	110.63	34.30
13.5	0	0.666	0.333	0.1518	−0.0478	110.63	34.30	75.19	24.81
15.5	0	0.666	0.333	0.0931	−0.0274	218.36	63.33	110.63	34.30
16.5	0	0.666	0.333	0.1856	−0.0585	110.63	34.30	75.19	24.81
18.5	0	0.666	0.333	0.1111	−0.0327	218.36	63.33	110.63	34.30
20	0	0.666	0.333	0.1201	−0.0353	218.36	63.33	110.63	34.30
21.5	0	0.666	0.333	0.1291	−0.0380	218.36	63.33	110.63	34.30
2	0	0.5	0.5	0.0133	−0.0040	218.36	63.33	110.63	34.30
4	0	0.5	0.5	0.0300	−0.0090	75.19	24.81	218.36	63.33
5.5	0	0.5	0.5	0.0412	−0.0124	75.19	24.81	218.36	63.33
7	0	0.5	0.5	0.0524	−0.0157	75.19	24.81	218.36	63.33

Table B2.
Part of the data used in training results (data2).

Inputs						Expected outputs			
Global Stress in x (σx) Pa	**Global Stress in y (σy) Pa**	**Volumetric fraction, material 1 (n1)**	**Volumetric fraction, material 2 (n2)**	**Strain in x (Ɛx)**	**Strain in y (Ɛy)**	**Q11 Material 1 (Q11M1) GPa**	**Q12 Material 1 (Q12M1) GPa**	**Q11 Material 2 (Q11M2) GPa**	**Q12 Material 2 (Q12M2) GPa**
1.84	0	0.429	0.571	0.0227	−0.0072	110.63	34.30	75.19	24.81
2.68	0	0.600	0.400	0.0167	−0.0049	218.36	63.33	110.63	34.30
3.52	0	0.429	0.571	0.0246	−0.0073	75.19	24.81	218.36	63.33
4.36	0	0.600	0.400	0.0502	−0.0159	110.63	34.30	75.19	24.81
5.20	0	0.429	0.571	0.0364	−0.0108	218.36	63.33	110.63	34.30
6.04	0	0.600	0.400	0.0502	−0.0153	75.19	24.81	218.36	63.33
6.88	0	0.429	0.571	0.0848	−0.0271	110.63	34.30	75.19	24.81
7.72	0	0.600	0.400	0.0482	−0.0142	218.36	63.33	110.63	34.30
8.56	0	0.429	0.571	0.0598	−0.0178	75.19	24.81	218.36	63.33
9.40	0	0.600	0.400	0.1083	−0.0342	110.63	34.30	75.19	24.81
10.24	0	0.429	0.571	0.0717	−0.0214	218.36	63.33	110.63	34.30
11.08	0	0.600	0.400	0.0921	−0.0280	75.19	24.81	218.36	63.33
11.92	0	0.429	0.571	0.1469	−0.0469	110.63	34.30	75.19	24.81
12.76	0	0.600	0.400	0.0797	−0.0235	218.36	63.33	110.63	34.30
13.60	0	0.429	0.571	0.0951	−0.0284	75.19	24.81	218.36	63.33

Table B3.
Part of the data used in validation results (data3).

Inputs						Expected outputs			
Global Stress in x (σx) Pa	Global Stress in y (σy) Pa	Volumetric fraction, material 1 (n1)	Volumetric fraction, material 2 (n2)	Strain in x (Ɛx)	Strain in y (Ɛy)	Q11 Material 1 (Q11M1) GPa	Q12 Material 1 (Q12M1) GPa	Q11 Material 2 (Q11M2) GPa	Q12 Material 2 (Q12M2) GPa
8.731	0.000	0.671	0.329	0.1135	−0.0380	75.19	24.81	110.63	34.30
1.867	0.000	0.662	0.338	0.0205	−0.0066	110.63	34.30	75.19	24.81
8.093	0.000	0.516	0.484	0.0940	−0.0310	75.19	24.81	122.32	40.37
8.731	0.000	0.329	0.671	0.1135	−0.0380	110.63	34.30	75.19	24.81
1.867	0.000	0.338	0.662	0.0205	−0.0066	75.19	24.81	110.63	34.30
8.093	0.000	0.484	0.516	0.0940	−0.0310	122.32	40.37	75.19	24.81
8.731	0.000	0.671	0.329	0.1135	−0.0380	75.19	24.81	110.63	34.30
1.867	0.000	0.662	0.338	0.0205	−0.0066	110.63	34.30	75.19	24.81
10.477	0.000	0.671	0.329	0.1362	−0.0456	75.19	24.81	110.63	34.30
2.240	0.000	0.662	0.338	0.0246	−0.0079	110.63	34.30	75.19	24.81
9.712	0.000	0.516	0.484	0.1128	−0.0372	75.19	24.81	122.32	40.37
10.477	0.000	0.329	0.671	0.1362	−0.0456	110.63	34.30	75.19	24.81
2.240	0.000	0.338	0.662	0.0246	−0.0079	75.19	24.81	110.63	34.30
9.712	0.000	0.484	0.516	0.1128	−0.0372	122.32	40.37	75.19	24.81
9.712	0.000	0.516	0.484	0.1128	−0.0372	75.19	24.81	122.32	40.37

Table B4.
Data used in the ANN application (data4).

Author details

Marta Eraña-Díaz and Mario Acosta-Flores*
Autonomous University of the State of Morelos, Cuernavaca, México

*Address all correspondence to: mario.acosta@uaem.mx

References

[1] Yang ZR. Machine Learning Approaches to Bioinformatics. Exeter, UK: World Scientific; 2010. p. 336

[2] Al-shayea QK. Artificial neural networks in medical diagnosis. International Journal of Computer Science Issues. 2011;**8**(2):150-154

[3] D'Antino T, Papanicolaou C. Mechanical characterization of textile reinforced inorganic-matrix composites. Composites Part B Engineering. 2017; **127**:78-91

[4] Abbud LH, Al-Masoudy MMM, Hussien Omran S, Abed AM. Experimental study the mechanical properties of nano composite materials by using multi-metallic nano powder/ epoxy. Materials Today: Proceedings. 2021. DOI: 10.1016/j.matpr.2021.06.395. ISSN 2214-7853

[5] Benyelloul K, Aourag H. Elastic constants of austenitic stainless steel: Investigation by the first-principles calculations and the artificial neural network approach. Computational Materials Science. 2013;**67**:353-358

[6] Shin HS, Lee SW, Kim CY, Bae GJ. Neural network based identification of nine elastic constants of an orthotropic material from a single structural test. In: Proceedings of the 21st ISARC; Jeju, South Korea; 2004

[7] Huang DZ, Xu K, Farhat C, Darve E. Learning constitutive relations from indirect observations using deep neural networks. Journal of Computational Physics. 2020;**416**:1-28

[8] Acosta-Flores M, Jiménez-López E, Chávez-Castillo M, Molina-Ocampo A, Delfín-Vázquez JJ, Rodríguez-Ramírez JA. Experimental method for obtaining the elastic properties of components of a laminated composite. Results in Physics. 2019;**12**:1500-1505

[9] Acosta-Flores M, Jiménez-López E, Rodríguez-Ramirez JA. Modelo para el análisis experimental de esfuerzos intralaminares en materiales compuestos laminados sujetos a carga axial. DYNA-Ingeniería e Industria. 2016;**91**:216-222

[10] Liu X, Tian S, Tao F, Yu W. A review of artificial neural networks in the constitutive modeling of composite materials. Composites Part B: Engineering. 2021;**224**:1-15

[11] Chen M, Challita U, Saad W, Yin C, Debbah M. Artificial neural networks-based machine learning for wireless networks: A tutorial. IEEE Communication Surveys and Tutorials. 2019;**21**(4):3039-3071

[12] Zhang Z. Artificial neural network. In: Multivariate Time Series Analysis in Climate and Environmental Research. Cham: Springer International Publishing; Springer, 2018. pp. 1-35

[13] Li X, Cheng X, Wu W, Wang Q, Tong Z, Zhang X, et al. Forecasting of bioaerosol concentration by a Back Propagation neural network model. Science of the Total Environment. 2020; **698**:134315

[14] Ye F, Wheeler C, Chen B, Hu J, Chen K, Chen W. Calibration and verification of DEM parameters for dynamic particle flow conditions using a backpropagation neural network. Advanced Powder Technology. 2019; **30**(2):292-301

[15] Srivastava N, Hinton G, Krizhevsky A, Sutskever I,

Salakhutdinov R. Dropout: a simple way to prevent neural networks from overfitting. The Journal of Machine Learning Research. 2014;**15**(1):1929-1958

[16] Li Z, Kamnitsas K, Glocker B. Overfitting of neural nets under class imbalance: Analysis and improvements for segmentation. In: Shen D, Liu T, Peters TM, Staib LH, Essert C, Zhou S, et al., editors. Medical Image Computing and Computer Assisted Intervention – MICCAI 2019. Cham: Springer International Publishing; 2019. pp. 402-410

[17] Frei S, Chatterji NS, Bartlett P. Benign overfitting without linearity: Neural network classifiers trained by gradient descent for noisy linear data. In: Proceedings of Thirty Fifth Conference on Learning Theory PMLR; 2022. p. 2668–2703

[18] Ying X. An overview of overfitting and its solutions. Journal of Physics: Conference Series. 2019;**1168**(2)

[19] Storey MA, Singer L, Cleary B, Figueira Filho F, Zagalsky A. The (r) evolution of social media in software engineering. Future of Software Engineering Proceedings. 2014:100-116. DOI: 10.1145/2593882.2593887

[20] Fritsch S, Guenther F, Guenther MF. Package 'neuralnet'. Training of. Neural Networks. The R Journal 2010;**2**(1): 30-38

[21] Riedmiller M, Braun H. Rprop: a fast adaptive learning algorithm. Procedures of the International Symposium on Computer and Information Science VII. 1992

[22] Riedmiller M, Braun H. A direct adaptive method for faster backpropagation learning: The Rprop algorithm. In IEEE International Conference on Neural Networks. 1993; p. 586-591

[23] Maplesoft, 2014, Maple (Release 18). Waterloo, ON: Maplesoft, a division of Waterloo Maple Inc.; 2014. Available online: https://hadoop.apache.org

Chapter 6

Optical Soliton Neural Networks

Eugenio Fazio, Alessandro Bile and Hamed Tari

Abstract

The chapter describes the realization of photonic integrated circuits based on photorefractive solitonic waveguides. In particular, it has been shown that X-junctions formed by soliton waveguides can learn information by switching their state. X junctions can perform both supervised and unsupervised learning. In doing so, complex networks of interconnected waveguides behave like a biological neural network, where information is stored as preferred trajectories within the network. In this way, it is possible to create "episodic" psycho-memories, able to memorize information bit-by-bit, and subsequently use it to recognize unknown data. Using optical systems, it is also possible to create more advanced dense optical networks, capable of recognizing keywords within information packets (procedural psycho-memory) and possibly comparing them with the stored data (semantic psycho-memory). In this chapter, we shall describe how Solitonic Neural Networks work, showing the close parallel between biological and optical systems.

Keywords: Nonlinear optics, photorefractive soliton, solitonic waveguide, supervised learning, unsupervised learning, Machine Learning, biological neural network, Artificial Intelligence, optical psycho-memory, optical neural network, photonics

1. Introduction

Software artificial intelligence (AI) and the neuromorphic approach, both electronic and optical, are born to reproduce the learning capacity of the biological neural system. AI software has now proved to be fundamental in many fields, although with the limits imposed by the tools used [1]. These represent the pretext for developing neuromorphic hardware capable of overcoming these limitations [2]. Neuromorphic optics has shown great versatility. However, current technologies reproduce only some aspects of neural biology without grasping the overall view. Works such as [3] implement fundamental units capable of reproducing excitability, or spiking properties, while others are focused on synaptic connections [4]. An overview is missing. The biology of the brain [5] teaches us that it is a system with local properties that can have global effects. In other words, learning is a process that affects entire regions of the neural network and manifests itself through a structural organization of the connections between neurons. In this way, real neural maps are built, whose development includes learning and memorization of information. Soliton neural networks (SNNs), exploiting the typical plasticity of photorefractive materials, are dynamic entities capable of self-modifying to process, learn and memorize information. Furthermore, they are able to do so selectively at the information level, exactly as it happens in the human brain. By physically combining the processing and memory units, SNN networks functionally approach the biological nervous system. Now learning and memorization are two events that occur at the same time through modifications of the spatial geometries.

2. Photorefractive solitons and solitonic waveguiding

2.1 Spatial solitons

The possibility of a beam becoming self-confining and propagating without diffraction was first studied about 50 years ago by E.T. Chiao, E. Garmire, and C.T. Townes [6] who in that year received the Nobel Prize for his studies on the maser and the laser. So they interpreted the phenomenon: "*We shall discuss here conditions under which an electromagnetic beam can produce its dielectric waveguide and propagate without spreading.*" Eight years later, V. E. Zakharov and A. B. Shabat formulated the theory of solitons [7].

The first experimental verification of self-confined beams arrived 13 years later, in 1985, by A. Bartelemy et al. [8] exploiting the Kerr-type nonlinearity of a liquid CS_2 cell and, 5 years later in 1990, within a glass planar waveguide [9].

It was immediately evident that the applicability of Kerr solitons was not simple: in fact, the low values of the nonlinearity of the excitable Kerr type in the glass required either very high intensity (GW/m^2) or very long propagations (being cumulative), and only planar geometries (Kerr solitons are stable only in 1D and not in 2D geometries). Over the years, it has been clear that these nonlinearities could be exploited only to realize temporal soliton behaviors (pulses without dispersion) in optical fibers by adopting long propagations but not within the chips.

However, in those years, and in particular in 1992–1996, the very first theoretical and experimental works on the formation of spatial solitons in photorefractive materials came out [10–21]. Only later on, at the beginning of the 2000s, bright solitons have been observed in lithium niobate (LN) [22] the most widely used nonlinear material for integrated devices. Since then, spatial solitons in LN have been largely used as waveguides in devices.

However, the first use of solitons as waveguides started early: in 1991, De la Fuente et al. [23] used Kerr solitons as waveguides. Almost 9 years later, E. Fazio et al repeated the same experiment in a glass chip [24] and used spatial soliton interaction for signal processing [25].

2.2 Theory of photorefractivity and solitons

A photorefractive crystal is typically a semiconductor that has a second-order nonlinearity of the electro-optical type, that is, the possibility of varying its refractive index as a function of an applied static electric field. Mathematically this can be represented in terms of the nonlinear polarization intensity vector:

$$\vec{P}(\omega) = \varepsilon_0 \left[\overleftrightarrow{\chi}^{(1)} \bullet \vec{E}(\omega) + \overleftrightarrow{\chi}^{(2)} : \vec{E}(0)\vec{E}(\omega) \right] \tag{1}$$

where $\vec{E}(\omega)$ represents the electric field associated with the light and $\vec{E}(0)$ the static one. Putting in evidence the light field we obtain

$$\vec{P}(\omega) = \varepsilon_0 \left[\overleftrightarrow{\chi}^{(1)} + \overleftrightarrow{\chi}^{(2)} \bullet \vec{E}(0) \right] \bullet \vec{E}(\omega) \tag{2}$$

which shows that the electric susceptibility, and consequently the dielectric tensor, gets a linear dependence from the static field:

$$\overleftrightarrow{\varepsilon} = \varepsilon_0\left[1 + \overleftrightarrow{\chi}^{(1)} + \overleftrightarrow{\chi}^{(2)} \bullet \vec{E}(0)\right] \tag{3}$$

for this reason, it is also called the “linear Pockels effect.” Typically, the refractive index of crystals is described by an ellipsoid of the type:

$$\frac{x^2}{{n_x}^2} + \frac{y^2}{{n_y}^2} + \frac{z^2}{{n_z}^2} = 1 \tag{4}$$

and, as a consequence, its variation is expressed by the variation of

$$\frac{1}{n^2}\ \text{(terms)}$$

$$\Delta\left(\frac{1}{n_i^2}\right) = \sum_j r_{ij}E_j(0) \tag{5}$$

that corresponds to a decrease of the refractive index:

$$n_i[E(0)] = n_{i,0} - \frac{1}{2}\sum_j n_{i0}^3 r_{ij}E_j(0) \tag{6}$$

where i represents one of the crystallographic directions (x,y,z) and n_{i0} describes the linear refractive index along the *i-th* direction.

There are two critical points in the discussion that has followed so far:

1. the local electrostatic field must give a local distribution to induce an electro-optical variation of the refractive index capable of self-confining the laser beam and originating spatial solitons

2. the electro-optical effect decreases the refractive index of the material (see Eq. (6)) while to self-confine the light a waveguide must have a higher refractive index than the surrounding environment.

For these reasons, it is necessary to follow a small procedure, a kind of small trick, to achieve a positive variation of the refractive index capable of self-confining the light: this can be done by applying a bias field to the whole material that lowers its index everywhere, and screening it in a small region where the light is, in order to raise back its value. As a consequence, the bright photorefractive spatial solitons are usually called *screening solitons*. Here is how this happens.

Let us consider a photorefractive medium as a semiconductor doped by a donor medium. Donor states (N_D) are usually localized energetically within the energy gap: which means that light can induce electron transitions from the donor states to the conduction bands. Consequently, two charge populations are generated so far: ionized donors, that is holes (N_D^+), which are physically localized, that is, are not free of moving because are connected to the physical position of the dopant ions, and electrons, which instead can go everywhere being in delocalized conduction states.

The donor rate equation is:

$$\frac{\partial n_D^+}{\partial t} = \sigma F n_D - \gamma n_d^+ n_e \tag{7}$$

where σ is the absorption cross-section, F the photon flux, and γ the relaxation rate. The electron rate equation follows the donor one, with the inclusion of the diffusion-conduction terms:

$$\frac{\partial n_e}{\partial t} = \frac{\partial n_D^+}{\partial t} - \mu \vec{\nabla} \bullet \left(n_e \vec{E} + \frac{k_B T}{q} \vec{\nabla} n_e \right) \tag{8}$$

where μ is the electron mobility, k_B the Boltzmann constant, and T the temperature. Electrons and holes constitute the local charge density ρ:

$$\rho = q\left(n_D^+ - n_e\right) \tag{9}$$

which generates, through Gauss's theorem, a local electric field that screens the applied bias:

$$\varepsilon \vec{\nabla} \bullet \vec{E}_{SC} = \rho \rightarrow \vec{E}_{local} = \vec{E}_{bias} + \vec{E}_{SC} \tag{10}$$

Applying a bias field along the extraordinary $\hat{c}$ the crystallographic direction of a uniaxial photorefractive crystal, the refractive indices get the expressions

$$\begin{cases} n_x = n_y = n_0 \\ n_z = n_e - \dfrac{1}{2} n_e^3 r_{33} E_{z-local} \end{cases} \tag{11}$$

The nonlinear light propagation is then described by the nonlinear wave equation [13]:

$$\left[\frac{\partial}{\partial x} - \frac{i}{2k}\left(\frac{\partial^2}{\partial y^2} + \frac{\partial^2}{\partial z^2}\right)\right] A(x, y, z) = \frac{ik}{n} \delta n(E_{local}) A(x, y, z) \tag{12}$$

where the field amplitude, in the case of a self-confined solitonic solution, should be factorized into an amplitude, independent from x, and a propagative term as follows:

$$A(x, y, z) = u(y, z) e^{i(\omega t - \Gamma x)} \tag{13}$$

as done for every kind of soliton, not only the photorefractive ones. Many groups have tried to solve analytically Eq. (12) without real success. Semi-analytical solutions are indeed reported in the literature showing that such complex problems can support bright solitons. In order to observe the soliton formation, a numerical integration (*FDTD—Finite Difference in Time Domain*) is performed of all Eqs. (6)–(11). Often, an approximated equation is considered, taking into account the saturable behavior of the nonlinear dielectric constant:

$$\left[\frac{\partial}{\partial x} - \frac{i}{2k}\left(\frac{\partial^2}{\partial y^2} + \frac{\partial^2}{\partial z^2}\right)\right] A = -\frac{\in_{NL} E_{bias}}{1 + \frac{|A|^2}{|A_{SAT}|^2}} A \tag{14}$$

2.3 Experiments on photorefractive solitons

The experimental set-up for spatial solitons is shown in **Figure 1** [22]. A laser beam (soliton beam) is focused down to about 10–12 μm FWHM onto the input face of a

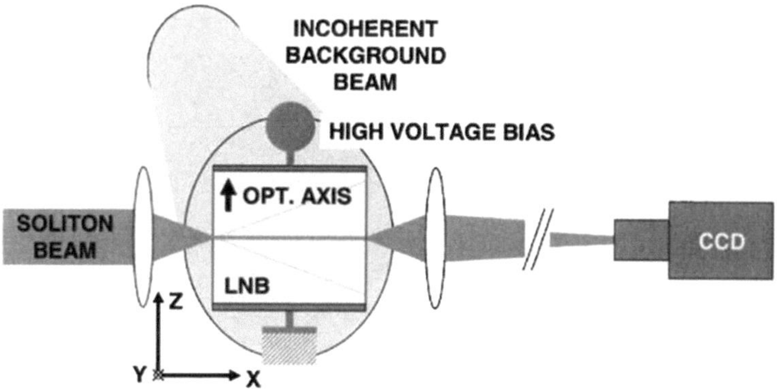

Figure 1.
Experimental set-up for screening photorefractive solitons [22].

sample. To generate suitable refractive index modulation, the sample must be biased along its optical axis.

The value of the electric field to bias strongly depends on the crystal type and its electro-optic coefficient: for example, using strontium barium niobate crystals (SBN) which have a very high electro-optic coefficient, the electric field ranges from a few hundred V/cm up to some kV/cm [26]; lithium niobate (LN) has a lower electro-optic coefficient and requires several tens of kV/cm [22]; in materials with high optical activity like $Bi_{12}SiO_{20}$ (BSO), the applied bias must be as high as 55 kV/cm or higher to induce the light to reach a nonlinear polarization regime and to self-confine [27–29]. Chauvet et al. [30, 31] proposed an interesting innovative solution for the bias application: induce an internal electric field by applying a thermal gradient and take advantage of the pyroelectric effects that some crystals have, for example, LN. Indeed, this is a major improvement in the technology, as it eliminates any conductive contacts/plates, thus leaving the sample completely free and accessible from all sides for further applications.

Background illumination can be provided also, to stabilize the solitonic beam during propagation (i.e., prevent beam self-deflection [32–35]).

Finally, an optical imaging system is placed after the sample to monitor the output face of the sample using a camera. The typical evolution of the soliton formation is shown in **Figure 2** where the light intensity at the output phase is shown.

A key feature of photorefractive solitons is the very low power required for their writing. Photorefractive solitons require very low powers, of the order of microwatt in continuous [36]. This means that they can be made both with coherent light from continuous or pulsed lasers at the fundamental or second harmonic frequency [37–40], even in the femtosecond regime [41, 42], and with incoherent light from fluorescent bulbs [43] or even ion fluorescence [44].

E. Fazio et al. [22] have shown experimentally that the solitonic solution gives a hyperbolic transverse profile which can be easily identified by plotting the transverse intensity distribution in a semi-log scale (**Figure 3**). A laser beam has usually a Gaussian profile that, in a semi-log plot, gets a negative parabolic shape. As soon as it evolves into a soliton, the Gaussian profile rearranges into a hyperbolic one. This transformation can be monitored in the semi-log graph, where the hyperbolic profile gets a triangular shape (a linear rise and fall tuned together on the vertex).

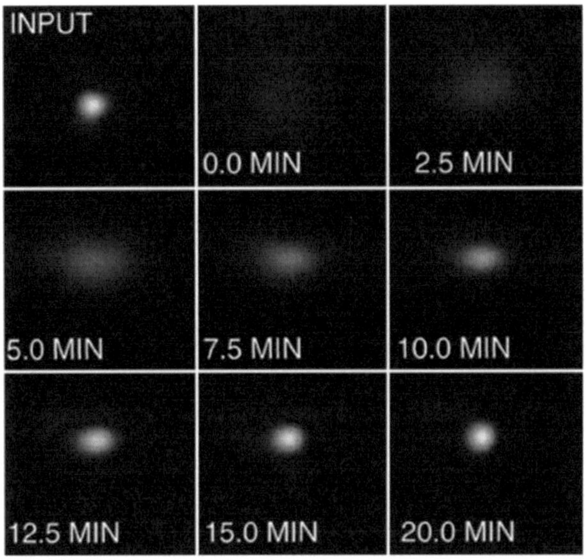

Figure 2.
Experimental images of the soliton formation and stabilization [22].

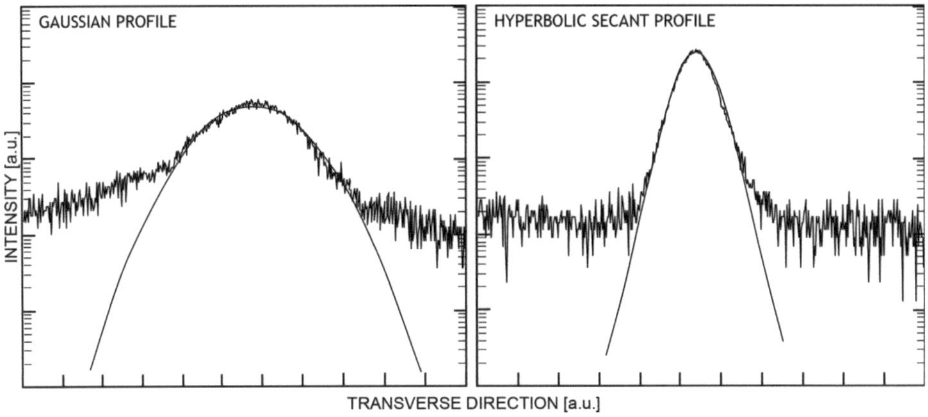

Figure 3.
A Gaussian laser beam modifies into a hyperbolic secant beam when becomes spatial solitons [22].

2.4 Photorefractive soliton waveguiding

Among the possible applications of soliton beams, one of the most important is their use as waveguides. Compared to traditional techniques, writing solitonic waveguides has many advantages in terms of construction costs, 3D geometries, propagation characteristics, and time durations.

Regarding costs, solitonic waveguides can be written with extremely low laser powers and above all in continuous mode: therefore, practically at no cost, since they can also be written by laser diodes for a few euros.

With regard to 3D geometries, a soliton guide can be written in any position within a nonlinear substrate, allowing full exploitation of the entire available volume. This

Writing powers	From nW up to mW (typically μW)
Full Width Half Maximum	Typically 10–18 μm (min 2–3 μm)
Propagation	Measured up to 2–3 cm
Refractive index contrast	Typically $10^{-3} \div 10^{-4}$
Refractive index profile	Hyperbolic secant—Gaussian
Waveguide modal dispersion	0.6 ± 0.2 fs/mm[1]
Lithium Niobate chromatic dispersion	9.9 ± 0.2 fs/mm[2]
Waveguide propagation losses	0.07–0.04 dB/cm
Waveguide lifetime	From few ns to months

[1]Measured at 800 nm with 75 fs pulses within waveguides written at 514 nm[2]Measured at 800 nm with a CW laser beam

Table 1.
Performances of typical photorefractive solitonic waveguides.

was not possible before with traditional waveguide construction techniques, which act mainly on the surface of the substrate or at most by penetrating a few microns.

Regarding the propagative characteristics, the performances of a solitonic guide are amazing, significantly improving the specifications of traditional waveguides.

Table 1 shows some characteristic values of a soliton guide made of lithium niobate. As you can see, the waveguides are relatively wide and with a rather low refractive index contrast. These factors are related to the applied bias electric field: Low fields originate wide beams with a modest contrast; very high fields can originate narrow beams and, consequently, high refractive index contrasts.

However, the fundamental characteristic of soliton waveguides is linked to the propagative losses, extremely low in the order of 0.07–0.04 dB/cm (limit of measurability), much lower than commercial waveguides (a guide obtained by ion exchange typically has 1 dB/cm as propagative losses). This factor is related to the nature of solitons: unlike traditional guides in which the index profile is made artificially, in this case, it is precisely the light that chooses the best index profile to be able to propagate self-confined, that is, without diffraction. This leads to ultra-very low losses and low modal dispersion (since the guides are almost single-mode).

Another fundamental characteristic of solitonic guides is their transient, permanent or semi-permanent character: using substrates with a very rapid dielectric relaxation and/or using thin films, as soon as the writing light is turned off the associated guide disappears, with times even of a few nanoseconds. By using substrates with extremely slow dielectric relaxations [45, 46], waveguides can survive for a long time, even months. When writing solitons with very intense femtosecond pulses, the material can undergo permanent changes and the waveguides no longer erase.

3. Stigmergy, reinforcement learning, and photorefractive plasticity

3.1 Stigmergy

Stigmergy was first proposed by French entomologist Pierre-Paul Grassè in the1950s when studying the activities of social insects [47]. The word Stigmergy is a combination of the Greek words ``stigma" (outstanding sign) and ``ergon" (work), signifying that some activities of agents are prompted by external traces, which themselves are generated by the agent's activities [48]. Stigmergy allowed Grassè to explain how insects with fractional intelligence, without obvious communications, can collaboratively engage in complex tasks, such as building a nest simply by following very naive rules. In general, the paradigm of social insect societies is a distributed system that, despite the lack of sophistication of their individuals, offers a highly structured social organization. For instance, as a result of this organization, ant colonies can carry out complex assignments that in some cases are beyond the capacities of a single ant [49]. A study of their behavior indicates that in the heart of their commotional random movements, there can be seen the trace of a series of behaviors that are driven by repeated stimulus-response cycles [50]. For example, when searching for food, ants initially explore the area surrounding their nest randomly and while moving, they leave a chemical pheromone trail on the ground (**Figure 4**). Once an ant finds a food source, it evaluates the quantity and the quality of the food and carries some of it back to the nest [52].

During the return trip, the quantity of pheromone that an ant leaves on the ground may depend on the quantity and quality of the food. The pheromone trails will guide other ants to the food source and subsequently, the shortest path to the food source will be reinforced as the result of a higher probability of feedback concerning the long paths [51]. This environment-intermediated type of communication has captivated researchers in many dissimilar fields. For example, it can be referred to all those protocols for the optimization of multi-variable problems known as genetic algorithms, which exploit the rules of genetics to solve mathematical problems with many independent variables, or neural networks, mathematical systems that base the calculation on a "learning" database that the system has previously prepared. All these typical problems that would require smart signal processing, are called "reinforcement learning" [53]. This expression is commonly used in computer science to describe those algorithms "of machine learning inspired by behaviorist psychology, which is

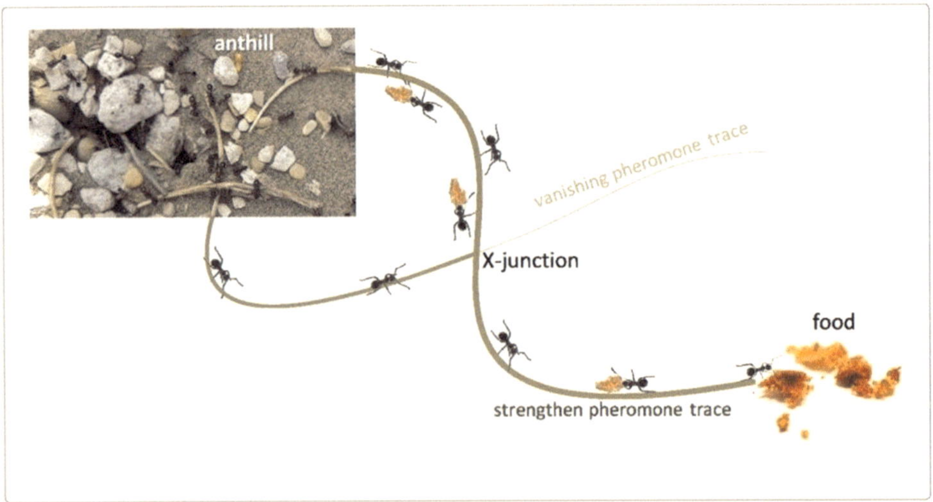

Figure 4.
Basic scheme of the search for food by the ants. The system is based on the two fundamental decision-making principles of following a trace of pheromone and of changing track when a more marked one is met [51].

connected with how software agents ought to take actions in an environment to amplify some impulse of cumulative reward [54].

3.2 Reinforcement learning

Reinforcement learning concerns neural networks or artificial intelligence protocols that self-set by reinforcing specific information identified by feedback in the system in order to solve complex problems. This procedure is indeed inspired by nature, adopting its Stigmergy in order to transfer information in decentralized systems, thus realizing distributed cognitive processes through many small, simple elaborations [55]. The basic idea of reinforcement learning is to consider the feedback derived from the dynamic interaction of the learning agent with the surrounding environment. Guiding autonomous agents to act optimally through trial-and-error interaction with the corresponding environment is the primary goal in the field of artificial intelligence and is regarded as one of the most important objectives of reinforcement learning [56]. During the learning process, the adaptive system tries some actions (i.e., output values) on its environment, then, it is reinforced by receiving a scalar evaluation (the reward) of its actions feedback [57]. As a result, the reinforcement learning algorithms selectively retain only the outputs that maximize the received reward because of the higher repetition rate over time [53].

Unfortunately, software-based protocols need solution times that increase exponentially with the size of the problem; after many years of research, no improved algorithm has been found to solve these problems within a polynomial time using a deterministic Turing machine. For this reason, hardware approaches have been proposed in the past [58, 59]. Among all, optical solutions to supercomputing seem to win for versatility [60] in terms of increased fan-in and fan-out, energy consumption, and recursive preprocessing. However, the proposed optical solutions [61, 62] neither reduce the complexity of the problem nor offer technologically efficient procedures without exponentially increasing the demand for physical resources [63].

Very recently, an alternative approach was proposed to realize photonic hardware able to perfectly simulate the Stigmergy processes adopted by ants searching for food. This alternative approach was published in the paper by M. Alonzo et al. entitled "All-Optical Reinforcement Learning in Solitonic X-Junctions [55]." In this work, the pheromone trajectories are represented by paths of the light through a nonlinear photorefractive material and the trajectory of the light is represented as the modified refractive index of the host material. Such modifications behave as induced waveguides, that is, regions that confine optical information which can travel inside them without being dispersed (as signals in optical fibers). The refractive contrast between the induced channel and the surrounding medium depends on the intensity of the writing beam. Consequently, it behaves like the pheromone quantity in the ant's path: it can be strengthened or weakened with the writing light intensity. This decision-making process can be represented by a nonlinear modulation of the crossing point between these paths. The strengthening of one path in an X-crossing point would correspond to making it a preferential trajectory, where the light will be conveyed more easily. It behaves as a channel of water whose banks have been made deeper and therefore more capacious: when two channels meet, more water will flow into the deepest channel rather than inside the shallowest one. Such addressable behavior has been induced into a nonlinear optical X-junction. The junction has been realized by injecting two absorbed beams that cross each other in the middle of the host photorefractive medium. Each beam modulated the refractive index of the host medium according to its intensity. A signal beam (unable to modify the host medium) is injected inside one channel and consequently reaches the X junction. It represents the information that propagates inside the photonic structure. If the writing beams have the same intensities, the junction is perfectly symmetrical, meaning that 50% of the information beam emerges from one channel and 50% from the other one (**Figure 5**). When the writing light is unbalanced or writing feedback is injected from the output, the X-junction switches to an asymmetric behavior, for which 80% of the information beam is now conveyed inside the strengthened channel and the remaining 20% remains in the weaker one.

3.3 Photorefractive plasticity

In neuroscience, this phenomenon is the basis of the selective memorizing-forgetting process that characterizes the memory of the events in the brain [64]: information pieces that are no longer reinforced will gradually be lost concerning recently reinforced ones. This capability arises owing to the considerable plasticity of the individual building block of the nervous system which allows animals to adapt to changing internal and external environments. During development, learning, and ongoing behavior, individual neurons, synapses, and circuits form short-term and long-term changes as a result of experience. This is the basis of the learning in a neural network which governs neuroplasticity, that is the ability of a system to modify the synaptic interconnection network according to its own needs, both to carry out "reasoning" and to recover unused areas (e.g., reusing regions that are inhibited due to trauma or injury) [65]. Neuroplasticity occurs at all levels, from the behavior of a single ionic channel to the morphology of neurons and large circuits and over

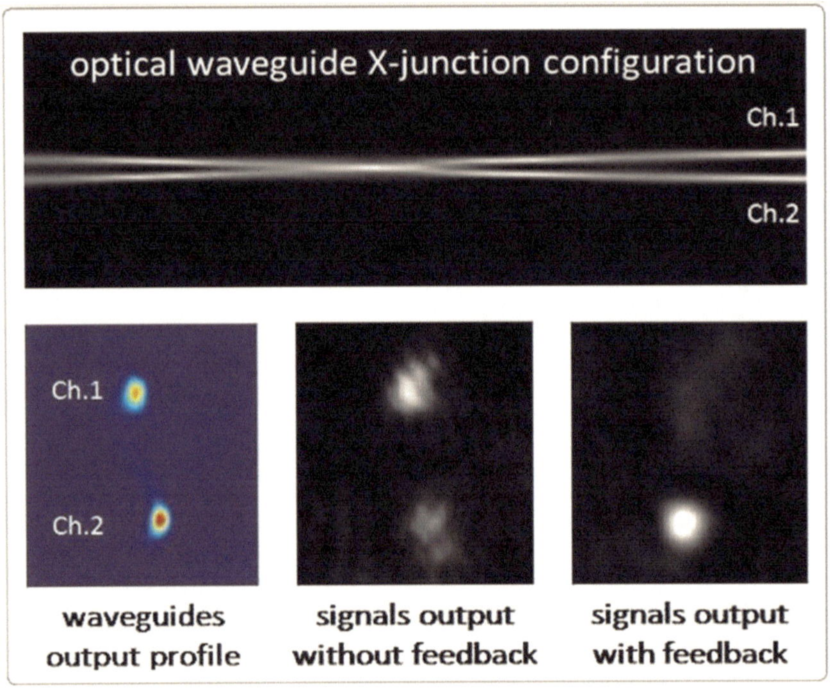

Figure 5.
Numerical simulation (top) and experimental results (below) of a stigmergic photonic x-junction [51].

timescales ranging from milliseconds to years [66]. Each level of the elementary unit is connected in parallel to each other, which performs simple operations of the storage and processing of information in successive cascading levels. Similar to the performance of the ants in the colony the information processed by a group of neurons is sent to the next level of neurons by opening and/or closing specific synaptic interconnections. In this way, the memory and subsequent reasoning consist of trajectories within the network, the mapping of which represents the set of stored information, which can be kept over time or deleted as needed. This is the way of "learning" and "remembering" a biological neural network. Any further information will follow its path: if the new path coincides with an active trajectory, then the information will be recognized, otherwise, the signal will be blocked, sooner or later, by inactive synaptic interconnections [67].

In this way, the neural network simultaneously remembers and processes, in a spatial coexistence where, the traditional computers cannot do this: in fact, they are based on the Von Neumann architecture which provides one or more processors connected to various external, separate peripherals, including memory. Whenever the computer needs information, it must access the memory to take and bring data back to the processor. This operation requires machine time and costs, in terms of energy consumption. Whereas, the neuromorphic paradigm, on the other hand, wants to unify the two areas of processing and memory, as happens in the biological field. However, overcoming the dichotomy between processing and memory is possible by creating neuromorphic architectures. By exploiting the typical functional geometries of the nervous system, information can be stored and processed in the same physical location, unifying memory and processor. In 2011, C. David Wright introduced the use of PCM for arithmetic and bio-inspired calculation [68], and provided the principle experimental proof of "processor" based on PCM for the first time, demonstrating the four basic operations of addition, multiplication, division and subtraction, and storing the results at the same time. In the same year, D. Kuzum reported new nanoscale electronic synapses based on PCM for optical data storage and non-volatile storage [69]. Continuous resistance transitions in PCM [70] and saturable absorber composite materials are used to simulate the properties of biological synapses so as to realize synaptic learning rules [71]. In 2017, Alexander N. Tait of Princeton University published a paper referred to neuromorphic silicon photonics, introducing the world's first integrated photonic neural network [72], It uses a neural compiler to program a silicon photonic neural network with 49 nodes, each node operates at a specific wavelength, light from each node is detected and summed before it is fed into the laser, then the output will be feedback to create a feedback loop with nonlinear characteristics. Tait et al. simulated traditional neural networks demonstrated how photonic neural networks can solve differential equations and found that photonic neural networks using silicon photonic platforms can be connected to ultrafast information processing environments for radio control and scientific computation.

It should be noted that most of the platforms that have been introduced as photonic or electronic neural networks are fixed structures that rigidly perform the calculation without the capability of changing the interconnections as requested [73]. This last aspect requires the use of modifiable plastic materials and/or devices, that is, capable of assuming different behaviors depending on the information to be stored. In these structures, the configuration of the neurons and their interconnections are written and predefined. So, they are only capable of doing certain limited functions, whereas the biological neurons can dynamically modify the interconnections in the procedure of training. They can establish new interconnections or if requires they can diminish or strengthen the weight of specific interconnections in synaptic points. Recently, inspired by the biological brain, reinforcement learning methods based on the memory of past experiences have been realized in photonic platforms via solitonic interconnections. Thanks to the plasticity of photorefractive materials, the light beam itself is able to locally vary the refractive index of the host material and create a channel within which it can propagate without diffraction. This solitonic signal will change the refractive index of the medium similar to the pheromone-mediated indirect communication. Obviously, the repetition and intensity profile of the incoming signals will affect the formation of the waveguide channel by exploiting the nonlinearity of the refractive index. This channel can also be used by other beams that recognize it as a waveguide. Also, these interconnections have a specific lifetime and their weight's strength is dependent on the extent of their exploitation. The interconnection's existence and strength is a self-driven process in which the signal itself can reconfigure its pathway by its occurrence redundancy. Consequently, any interconnection which is not activated for a long time will be diminished and taken out of the computation cycle, at the expense of the highly exploited ones. Depending on the material used, this waveguide will then cancel itself completely when the writing light is switched off (rapid dielectric relaxation) or survive for a shorter or longer time (slow dielectric relaxation).

The solitonic guides are, therefore, completely plastic guides, which are induced by a modification of the material and can be suitably shaped by other light passing through them. Now there is no artificial neuroplastic hardware, that its networks be able to reorganize themselves autonomously, although this is the only way to reproduce artificial systems similar to biological ones. An extremely promising way to achieve them is represented by soliton optical neural networks, able to exploit the plasticity of the refractive index to create circuits whose interconnections can be activated or inhibited as required by the information to be stored or processed. In 2018, a collaboration between Sapienza and Nanyang Technological University in Singapore demonstrated that X-junctions formed by soliton waveguides learn information [55]. Recently, it has been shown that X-junctions can perform both supervised and unsupervised learning, behaving as if they were neurons that fully exploit the plasticity of the substrate both to write the circuit and to post-modification based on the evolution of the system [74]. By exploiting the X junctions as elementary units, it is possible to create complex neural networks capable of storing information as specific trajectories within the circuit network [75].

4. Solitonic X-junctions as photonic neurons: Supervised and unsupervised learning

The solitonic neuron is a device capable of reproducing the fundamental characteristics of the learning and memorization processes typical of biological neurons. From a biological point of view, the neuron, a fundamental unit of the nervous system, is a dynamic unit capable of self-assembling and self-modifying according to the information that arrives. These structural changes are the mirror of the unfolding of learning and memorization [76]. The capacity for self-organization is not local, that is, it does not affect the individual units independently as if they were non-communicating structures. Whenever a certain type of information presents itself at the gates of the nervous system, through the different types of receptors of which it is composed, an enlarged (global) mechanism is set in motion, influencing pathways within the nerve mapping, affecting neurons through connections both in parallel and in series. This characteristic interconnectedness underlies the functional complexity of the nervous system and at the same time represents its strength. This is why an event, at a precise point on the neural map, can trigger a succession of changes culminating in a complete reorganization of entire neural regions. One of the properties we have already talked about previously but that needs to be brought to attention perhaps with greater energy is the concept of plasticity. This is a key feature for various reasons. A good approximation of the concept of plasticity can be defined through the expression "dynamic self-organization" [44]. It is typical of systems that do not remain identical to themselves neither at the functional nor at the structural level [77]. More precisely plastic hardware overlaps these two nuances: function becomes synonymous with structure. This is one of the fundamental properties of biological neural tissue. Conceiving the implementation of an artificial hardware neuron that works on the biological model, there are therefore some characteristics that should be kept in mind. First of all, it must have a dynamic structure that is able to adapt to the evolution of the environment and provide a response to it in a nonlinear way. Furthermore, it is necessary to keep the chronology of the information processed at the same time as the analysis and learning operations. A schematic of the functional blocks which characterize the "*modus conoscendi*" of a neuron is shown in **Figure 6a**.

We can highlight a tripartite structure: the neuron receives signals through the dendrites, small branches acting as input channels. The information is collected through these and conducted to the soma, the central body of the neuron which acts as a real microprocessor. Here the signals are "read" and analyzed through weighing and comparison operations with respect to a threshold value. A signal above the threshold is highly informative so it must be stored and propagated along the neural mapping. On the contrary, a sub-threshold signal is judged not important at the informational level and, therefore, its transmission is stopped [78]. The axon is a long channel with the task of carrying the signal out and distributing it to the neurons that follow through special connections called synapses. These are the basis of the communication between different units and correspond to the entities that allow the realization of complex neural mapping. The soliton photon neuron, which the research group of the

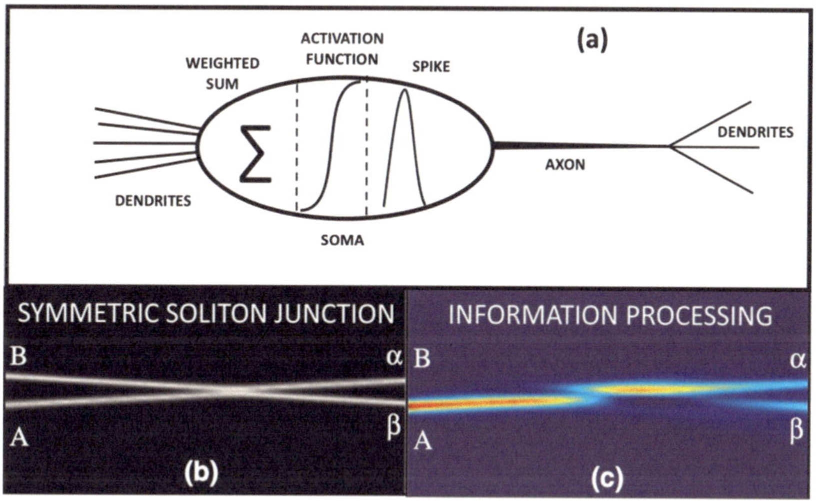

Figure 6.
(a) Fundamental scheme of a biological neuron. (b) the solitonic neuron X-junction structure. (c) Perfectly balanced X-junction.

Smart and Neuro Photonics Lab has designed and built, has a functional geometry very close to the one just described. A soliton neuron [55] is characterized by an X-junction structure [79], as shown in **Figure 6b** obtained through the intersection of self-written waveguides by two self-confining and non-diffracting laser beams. Using the technology of spatial solitons obtained through the Pockels effect [13, 80], the writing takes place through a local variation of the refractive index induced by the incoherent laser light beams. All materials with a saturating nonlinear electro-optical coefficient can be used. The input channels functionally represent the dendrites that collect the input signals. The soliton soma coincides with the region in which the two laser beams progressively approach until they overlap. It is in this region, which by assonance with the ML models we call the solitonic node, that the nonlinear energy transfer between the channels takes place which, as we will see shortly, allows the learning process. The output channels, which allow a subsequent redistribution of the propagated signal, replicate the functional action of the axon. In order for the soliton soma to form and be active at a functional level, it is necessary that the laser beams arrive at the input face of the crystal at an extremely small angle with respect to the normal, between 0.8° and 1°. For different angles, the node is characterized by an area that is too limited which determines a low coupling between the waveguides. The soliton neuron can perform supervised and unsupervised learning tasks [55, 74]. From a theoretical point of view, supervised learning is performed using a fundamental truth, or in other words, there is prior knowledge of what the output values to learn should be [81–83]. If the learning is unsupervised, on the contrary, there is no a priori knowledge of the desired output, which is identified at the same time as learning [84, 85]. The substantial difference lies in the way in which the already written waveguide structure is modified. In the supervised case, indeed, it is necessary to know the target and therefore to guide the learning. The X junction is modified using a feedback system that locally alters the refractive index contrast, depending on the information received, through successive cycles (**Figure** 7). This mechanism is fully

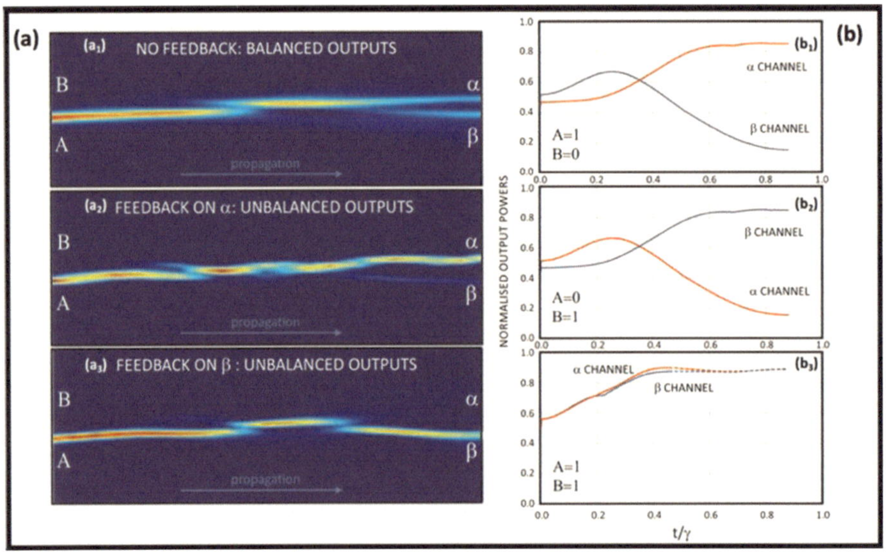

Figure 7.
The X-junction neuron switches from the balanced outputs (a) to the unbalanced behaviors, either due to feedback on the alpha channel (b) or due to feedback on the beta channel (c). Learning dynamics of the solitonic junction: starting from the initial neutral condition 50/50, the junction recognizes the input and switches accordingly.

explained in [74], where a numerical code FDTD solving the nonlinear equation (Eq. (1)) reported below, shows the morphological evolution of the neuron (see **Figure** 7), index of learning what is happening.

$$\nabla^2 A_i = -\frac{\in_{NL} E_{bias}}{1 + \frac{|A_1|^2 + |A_2|^2}{|A_{SAT}|^2}} A_i \tag{15}$$

where ε_{NL} is the nonlinear dielectric constant, E_{bias} is the electrostatic bias field that allows the formation of photorefractive solitons and $|A_{SAT}|^2$ the saturation intensity. In this type of learning, only the A_1 and A_2 beams are able to excite the nonlinearity underlying the index modification. The information signal is indeed represented by a laser with a different wavelength, with respect to which the refractive index is not sensitive. The initial situation is represented by a perfectly balanced X junction, as shown in **Figure 7a₁**, which is characterized by a symmetrical structure obtained by using two laser beams at the same power input. The injected signal, having reached the solitonic node, "perceives" the same index and divides itself perfectly 50% into the two output channels. By using different power ratios in the writing phase, it is possible to build asymmetrical structures **Figure 7a₁** and **7b₃**. In this case, the index will begin to differentiate already within the area of the soliton soma and will result in an unequal division of the input information in the two outputs. However, the soliton neuron is also able to perform unsupervised learning tasks. In this case, the refractive index of the crystal is also sensitive to the wavelength of the signal, which, by propagating within the previously written structure, is able to change it. The information becomes directly responsible for the asymmetrization of the junction. For unsupervised learning, the Helmholtz equation becomes:

$$\nabla^2 A_i = -\frac{\in_{NL} E_{bias}}{1 + \frac{|A_1|^2 + |A_2|^2 + \eta |A_3|^4 \left(1 - e^{-\frac{t}{\tau}}\right)}{|A_{SAT}|^2}} A_i \tag{16}$$

where A_3 represents the information signal and η an efficiency coefficient for the nonlinear process that depends on the wavelength and the material used. These structural variations can be the result of numerous successive propagation cycles or single events characterized by much higher powers. This is another point of similarity with the biological case. The biological signal, called spike, is propagated toward the axon when the combination of input signals is above the threshold. This can occur as a consequence of the accumulation of numerous inputs, spike trains, in a limited time interval, or by virtue of a very intense signal.

In the solitonic case, learning is, therefore, identified with the process of changing the refractive index and therefore has its own physical translation. What about memory? Many neuromorphic implementations, both in electronics and in optics, have achieved remarkable results in the reproduction of a neural system, however, there is always a great difficulty in defining a memory that is present at the same time as the processing unit. The soliton X junction introduces a new paradigm in the field of neuromorphic research, approaching the nature of biological neurons. The index modification is in general a semi-permanent property with times that depend on the particular material used. The input information is therefore saved in the particular morphological structure obtained during the learning phase. In, the authors show the possibility of building soliton neurons in bulk $LiNbO_3$ crystals. This represents the first supervised realization. The neuron is able to convey information, represented by a signal with a different wavelength, traveling within the waveguides in the directions declared by the local refractive index. Starting from these results and integrating them with the technology of spatial solitons in thin films in lithium niobate [86], in [87] the possibility of implementing soliton neurons in 8 μm films of lithium niobate is demonstrated. This technology brings with it numerous new benefits. First of all, its extreme compactness makes them a useful tool for integration into small devices. Furthermore, the Lithium Niobate layers show focusing dynamics of two orders of magnitude faster than the bulk counterparts. Finally, the films offer greater control over the propagation of the beams within the crystal, ensuring considerable precision which results in greater coupling and, ultimately, in a more performing soliton soma. By virtue of their plastic behavior, soliton X-junction neurons can be interfaced in more complex structures, to give rise to complex neural mappings capable of functionally replicating biological neural tissue. This perspective represents the great innovation of the soliton neuromorphic, which is not limited to reproducing a unity or a connection, as in previous neuromorphic models, but is able to reach a higher and more complete level of complexity, through the realization of a whole neural environment.

5. Bit-to-bit data storage and recognition

Solitonic neurons can be interconnected to form complex neural maps, and soliton neural networks (SNNs) [75]. Their functioning is based on the movement of photogenerated electrical charges that assume the same role played by neurotransmitters in biological neural networks (BNN). Both regulate the intensity with which a synaptic connection, solitonic or biological, is built, modified, or destroyed. Furthermore, the solitonic synapse, exactly as in the biological case, is the basis of the memorization processes. The repetition of information results in synaptic strengthening, which is synonymous with information memorization [88]. Therefore, learning and memorization are processes that occur through structural changes. In **Figure 8**, the summary diagram of the functionality of the BNNs and SNNs.

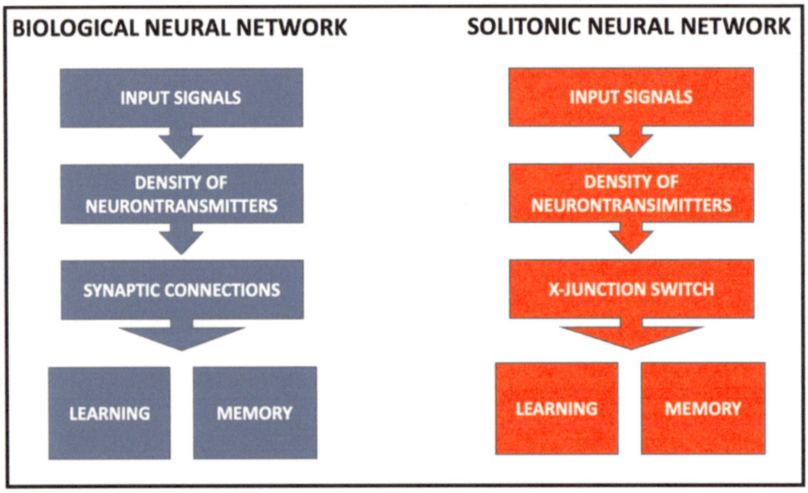

Figure 8.
Functional diagrams on the left of a BNN network and on the right of an SNN network. Both are able to self-modify their structure according to the information signals received to process and store them in precise neural patterns [75].

Recently, an SNN has been studied which is able to carry out a 4-bit recognition. It is formed by X-Junction channels written with equal power beams in order to create 50–50 junctions. SNNs are divided into successive layers as reported in **Figure 9**.

The first layer corresponds to the input face and is characterized by a number N of channels corresponding to the number of information (bits) to be processed and to the number of incoming laser sources. The SNN exploits the phenomenon of total reflection at the edge in correspondence with the even layers, which are therefore characterized by N/2–2 neural units. While the odd layers are characterized by N/2 fundamental units.

This network is able to learn by switching the propagating signal between the two outputs of each X-Junction. By appropriately increasing the size of the matrix it is possible to obtain the representation of any SNN. Each channel, therefore, has its own weight which is modified over time based on the information received as reported in equation.

$$\underline{Y} = W\left(E_{BIAS}, \underline{X}\right) \cdot \underline{X} \tag{17}$$

For an in-depth analysis of the SNN, we recommend reading [75].

An SNN network, at present, is able to perform an Episodic recognition. This term derives from psychology studies that have made it possible to identify three ways of working with memory, episodic, procedural, and semantic [89]. Memory is of an episodic type if it records an event photographically, that is, it fails to decontextualize the subjects present [90]. Let us consider the picture of a dog running in the mountains. The dog subject is recognized only in that environment, mountains, and in that position, running, if moved then it will be identified as different. Procedural memory, on the other hand, identifies a mechanism and learns its rule. Finally, semantic memory contains these mechanisms within itself, thus reaching abstraction through the analysis of details.

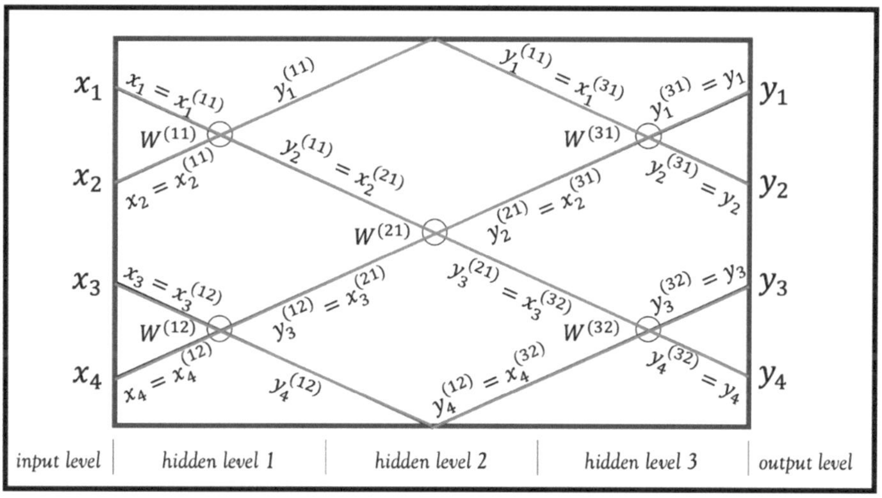

Figure 9.
Structure of a 4-bit SNN network. W is the weight of the junction point. In particular, $W^{(1)}$ is the weight relative to the node of the first solitonic neuron in layer 1, $W^{(2)}$ is the weight relative to the node of the second solitonic neuron in layer 2, and so on. The information inputs are represented by x_i while y_i the processed signals [75].

The solitonic technology has allowed, until now, to successfully realize an episodic memory able to save information through precise neural mapping. As information flows into the SNN it modifies the refractive index of the network, determining precise paths.

Learning the SNN takes place in two stages. The first phase, defined as Training, consists in administering the information pattern to be learned to the network several times. The network changes morphology accordingly. Then we try to understand how profound the changes have been, or how much the information has been learned and memorized. This phase is called validation. The network acts as a filter letting only the saved information propagate.

Figure 10 was realized starting from the results proposed by [75]. It shows the learning of 1-bit in four different cases corresponding to the four input channels. In **Figure 10a**, the first line shows the network training with the four 1-digit in each channel while the others are set to 0. The images below report the SNN recognizing process: it uses the stored information to operate the comparison. Therefore, if digit 1

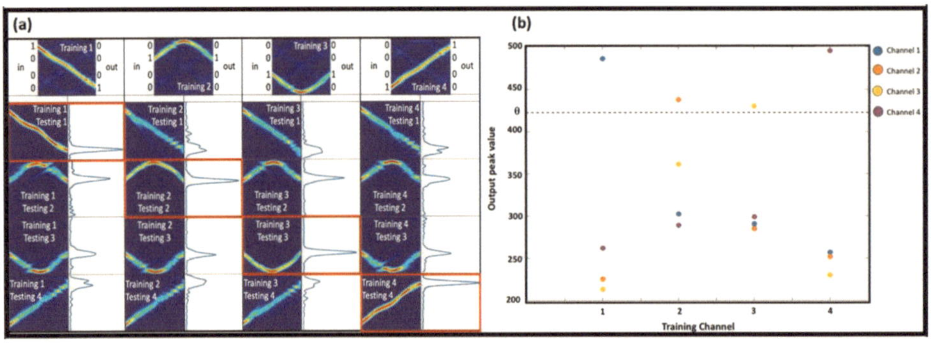

Figure 10.
In (a) training and validation processes of a 4-bit SNN are reported in 1-digit recognition case. The first line is related to the training phase while in the following rows validation steps are reported. In (b) the signal output amplitudes for different training numbers are reported: Only the trained channel is above the threshold (dotted line) [75].

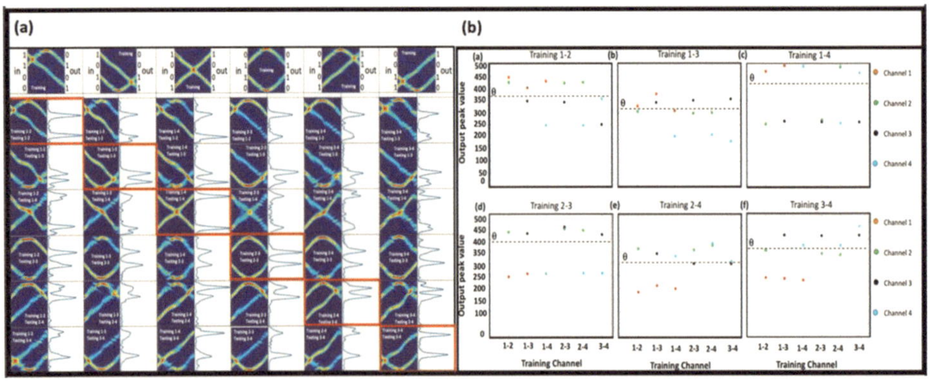

Figure 11.
In (a) training and validation processes of a 4-bit SNN are reported in 2-digit recognition case. The first line is related to the training phase while in the following rows validation steps are reported. In (b) the signal output amplitudes for different training numbers are reported: Only the trained channel is above the threshold (dotted line) [75].

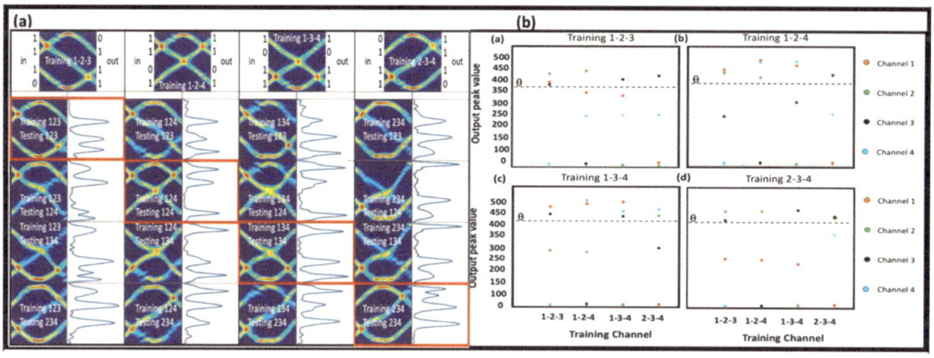

Figure 12.
In (a) training and validation processes of a 4-bit SNN are reported in 3-digit recognition case. The first line is related to the training phase while in the following rows validation steps are reported. In (b) the signal output amplitudes for different training numbers are reported: Only the trained channel is above the threshold (dotted line) [75].

of the new number corresponds to digit 1 of the training number, the output of the network is high and the information is recognized. Otherwise, the output is low, which means no recognition. **Figures 11** and **12** report the learning cases of 2-digit and 3-digit, following the scheme already illustrated in **Figure 10**.

The SNN recognizes through a threshold process. If the output is higher than a threshold, determined experimentally, then the recognition has occurred. This procedure can be generalized to N bits according to the equation shown below.

$$I_k^{output_i} \geq \theta I_k^{input_i} \tag{18}$$

where θ is the pure number (∼0.7).

Therefore, Optical Soliton Neural Networks are systems characterized by a structural dynamism, based on the plasticity of the refractive index, which can self-modify to recognize previously learned or new signals. Learning and memorization occur at the same time as physical evolutions.

6. Conclusions

Artificial intelligence is marking a profound innovation in everyday life. To overcome the limitations of AI software, research has developed the neuromorphic approach, which consists in reproducing the functional blocks of the human brain. A first attempt was carried out by electronics, which however suffer from a structural rigidity that does not match neural geometries. One of the fundamental qualities that characterize them is in fact plasticity, that is to say, the ability to self-modify one's units to trap, learning, and memory, in its structure. The solitonic optical approach that we have described in this chapter bases its effectiveness precisely on the concept of plasticity and self-assembly. Compared to other optical technologies, which focus on single neural properties (first of all excitability), soliton networks are able to reproduce complex behavior by exploiting the local differences in refractive indices to build specific trajectories for each information through the propagation of solitons. SNNs are currently able to reproduce a specific type of psycho-memory, episodic memory, in a particularly effective way, that is, with small powers (nW-μW) and with extremely low losses. SNNs capable of reproducing procedural and semantic memories are currently being studied. Once these objectives have been achieved, hardware that is functionally very close to biological neuronal dynamics will be available. In the biological neural system, the synaptic connections are created and deleted following the change in neurotransmitter density, in the soliton paradigm that we propose, the birth and modification of X-junction neurons depends on the density of photo-excited electric charges.

Acknowledgements

The authors would like to acknowledge all alumni that have been working previously in the Ultrafast Photonics Lab and then in the Smart and Neuro Photonics of Sapienza Università di Roma: all of them participated in increasing the in-depth knowledge we now have on photorefractive spatial solitons and their enormous application possibilities.

Conflict of interest

The authors declare no conflict of interest declaration.

Author details

Eugenio Fazio*, Alessandro Bile and Hamed Tari
Smart and Neuro Photonics Laboratory, Department of Fundamental and Applied Sciences for Engineering, Sapienza Università di Roma, Roma, Italy

*Address all correspondence to: eugenio.faziouniroma1.it

References

[1] Feldmann J, Youngblood N, Wright CD, Bhaskaran H, Pernice WH. All-optical spiking neurosynaptic networks with self-learning capabilities. Nature. 2019;**569**(7755):208-214

[2] De Lima TF, Shastri BJ, Tait AN, Nahmias MA, Prucnal PR. Progress in neuromorphic photonics. Nano. 2017; **6**(3):577-599

[3] Inagaki T, Inaba K, Leleu T, Honjo T, Ikuta T, Enbutsu K, et al. Collective and synchronous dynamics of photonic spiking neurons. Nature Communications. 2021;**12**(1):1-8

[4] Geng X, Hu L, Zhuge F, Wei X. Retina-inspired two-terminal optoelectronic Neuromorphic devices with light-tunable short-term plasticity for self-adjusting sensing. Advanced Intelligent Systems. Jun 2022;**4**(6):2200019

[5] Gerstner W, Kistler WM, Naud R, Paninski L. Neuronal Dynamics: From Single Neurons to Networks and Models of Recognition. Cambridge: Cambridge University Press; 2014

[6] Chiao RY, Garmire E, Townes CH. Self-trapping of optical beams. Physical Review Letters. 1964;**13**(15):479

[7] Shabat A, Zakharov V. Exact theory of two-dimensional self-focusing and one-dimensional self-modulation of waves in nonlinear media. Soviet Physics JETP. 1972;**34**(1):62

[8] Barthelemy A, Maneuf S, Froehly C. Propagation soliton et auto-confinement de faisceaux laser par non linearité optique de Kerr. Optics Communications. 1985;**55**(3):201-206

[9] Aitchison JS, Weiner A, Silberberg Y, Oliver M, Jackel J, Leaird D, et al. Observation of spatial optical solitons in a nonlinear glass waveguide. Optics Letters. 1990;**15**(9):471-473

[10] Segev M, Crosignani B, Yariv A, Fischer B. Spatial solitons in photorefractive media. Physical Review Letters. 1992;**68**(7):923

[11] Crosignani B, Segev M, Engin D, Di Porto P, Yariv A, Salamo G. Self-trapping of optical beams in photorefractive media. JOSA B. 1993; **10**(3):446-453

[12] Valley GC, Segev M, Crosignani B, Yariv A, Fejer M, Bashaw M. Dark and bright photovoltaic spatial solitons. Physical Review A. 1994;**50**(6):R4457

[13] Segev M, Valley GC, Crosignani B, Diporto P, Yariv A. Steady-state spatial screening solitons in photorefractive materials with external applied field. Physical Review Letters. 1994;**73**(24): 3211

[14] Zozulya A, Saffman M, Anderson D. Propagation of light beams in photorefractive media: Fanning, self-bending, and formation of self-pumped four-wave-mixing phase conjugation geometries. Physical Review Letters. 1994;**73**(6):818

[15] Shih M-F, Segev M, Valley G, Salamo G, Crosignani B, Di Porto P. Observation of two-dimensional steady-state photorefractive screening solitons. Electronics Letters. 1995;**31**(10): 826-827

[16] Zozulya AA, Anderson DZ. Propagation of an optical beam in a photorefractive medium in the presence of a photogalvanic nonlinearity or an externally applied electric field. Physical Review A. 1995;**51**(2):1520

[17] Zozulya A, Anderson D. Nonstationary self-focusing in photorefractive media. Optics Letters. 1995;**20**(8):837-839

[18] Zozulya A, Anderson D. Spatial structure of light and a nonlinear refractive index generated by fanning in photorefractive media. Physical Review A. 1995;**52**(1):878

[19] Fressengeas N, Maufoy J, Kugel G. Temporal behavior of bidimensional photorefractive bright spatial solitons. Physical Review E. 1996;**54**(6): 6866

[20] Segev M, Shih M-F, Valley GC. Photorefractive screening solitons of high and low intensity. JOSA B. 1996; **13**(4):706-718

[21] Shih M-F, Leach P, Segev M, Garrett MH, Salamo G, Valley GC. Two-dimensional steady-state photorefractive screening solitons. Optics Letters. 1996; **21**(5):324-326

[22] Fazio E, Renzi F, Rinaldi R, Bertolotti M, Chauvet M, Ramadan W, et al. Screening-photovoltaic bright solitons in lithium niobate and associated single-mode waveguides. Applied Physics Letters. 2004;**85**(12):2193-2195

[23] De La Fuente R, Barthelemy A, Froehly C. Spatial-soliton-induced guided waves in a homogeneous nonlinear Kerr medium. Optics Letters. 1991;**16**(11):793-795

[24] Fazio E, Zitelli M, Bertolotti M, Carrera A, Sanvito N, Chiaretti G. Solitonic waveguiding in planar glass structures. Optics Communications. 2000;**185**(4–6):331-336

[25] Zitelli M, Fazio E, Bertolotti M. All-optical NOR gate based on the interaction between cosine-shaped input beams of orthogonal polarization. JOSA B. 1999;**16**(2):214-218

[26] Duree GC Jr, Shultz JL, Salamo GJ, Segev M, Yariv A, Crosignani B, et al. Observation of self-trapping of an optical beam due to the photorefractive effect. Physical Review Letters. 1993; **71**(4):533

[27] Fazio E, Mariani F, Bertolotti M, Babin V, Vlad V. Experimental demonstration of (1+ 1) D self-confinement and breathing soliton-like propagation in photorefractive crystals with strong optical activity. Journal of Optics A: Pure and Applied Optics. 2001; **3**(6):466

[28] Fazio E, Ramadan W, Belardini A, Bosco A, Bertolotti M, Petris A, et al. (2+ 1)-dimensional soliton formation in photorefractive Bi 12 SiO 20 crystals. Physical Review E. 2003;**67**(2): 026611

[29] Fazio E, Ramadan W, Bertolotti M, Petris A, Vlad V. Complete characterization of (2+ 1) D soliton formation in photorefractive crystals with strong optical activity. Journal of Optics A: Pure and Applied Optics. 2003; **5**(5):S119

[30] Safioui J, Devaux F, Chauvet M. Pyroliton: Pyroelectric spatial soliton. Optics Express. 2009;**17**(24): 22209-22216

[31] Safioui J, Fazio E, Devaux F, Chauvet M. Surface-wave pyroelectric photorefractive solitons. Optics Letters. 2010;**35**(8):1254-1256

[32] Barthélémy A, Froehly C, Maneuf S, Reynaud E. Experimental observation of beams' self-deflection appearing with two-dimensional spatial soliton propagation in bulk Kerr material. Optics Letters. 1992;**17**(12):844-846

[33] Carvalho M, Singh S, Christodoulides D. Self-deflection of steady-state bright spatial solitons in biased photorefractive crystals. Optics Communications. 1995;**120**(5–6): 311-315

[34] Jinsong L, Keqing L. Screening-photovoltaic spatial solitons in biased photovoltaic–photorefractive crystals and their self-deflection. JOSA B. 1999; **16**(4):550-555

[35] Chauvet M, Coda V, Maillotte H, Fazio E, Salamo G. Large self-deflection of soliton beams in LiNbO 3. Optics Letters. 2005;**30**(15):1977-1979

[36] Fazio E, Ramadan W, Petris A, Chauvet M, Bosco A, Vlad V, et al. Writing single-mode waveguides in lithium niobate by ultra-low intensity solitons. Applied Surface Science. 2005; **248**(1–4):97-102

[37] Pettazzi F, Coda V, Chauvet M, Fazio E. Frequency-doubling in self-induced waveguides in lithium niobate. Optics Communications. 2007;**272**(1): 238-241

[38] Fazio E, Pettazzi F, Centini M, Chauvet M, Belardini A, Alonzo M, et al. Complete spatial and temporal locking in phase-mismatched second-harmonic generation. Optics Express. 2009;**17**(5): 3141-3147

[39] Fazio E, Belardini A, Alonzo M, Centini M, Chauvet M, Devaux F, et al. Observation of photorefractive simultons in lithium niobate. Optics Express. 2010;**18**(8):7972-7981

[40] Pettazzi F, Coda V, Fanjoux G, Chauvet M, Fazio E. Dynamics of second-harmonic generation in a photovoltaic photorefractive quadratic medium. JOSA B. 2010; **27**(1):1-9

[41] Vlad V, Petris A, Bosco A, Fazio E, Bertolotti M. 3D-soliton waveguides in lithium niobate for femtosecond light pulses. Journal of Optics A: Pure and Applied Optics. 2006;**8**(7):S477

[42] Pettazzi F, Alonzo M, Centini M, Petris A, Vlad VI, Chauvet M, et al. Self-trapping of low-energy infrared femtosecond beams in lithium niobate. Physical Review A. 2007;**76**(6):063818

[43] Mitchell M, Segev M. Self-trapping of incoherent white light. Nature. 1997; **387**(6636):880-883

[44] Fazio E, Alonzo M, Devaux F, Toncelli A, Argiolas N, Bazzan M, et al. Luminescence-induced photorefractive spatial solitons. Applied Physics Letters. 2010;**96**(9):091107

[45] Chauvet M, Hawkins S, Salamo GJ, Segev M, Bliss D, Bryant G. Self-trapping of planar optical beams by use of the photorefractive effect in InP: Fe. Optics Letters. 1996;**21**(17):1333-1335

[46] Alonzo M, Dan C, Wolfersberger D, Fazio E. Coherent collisions of infrared self-trapped beams in photorefractive. Applied Physics Letters. 2010;**96**(12): 121111

[47] Grassé P-P. La reconstruction du nid et les coordinations interindividuelles chezBellicositermes natalensis etCubitermes sp. la théorie de la stigmergie: Essai d'interprétation du comportement des termites constructeurs. Insectes Sociaux. 1959; **6**(1):41-80

[48] Theraulaz G, Bonabeau E. Coordination in distributed building. Science. 1995;**269**(5224):686-688

[49] Moffett MW, Garnier S, Eisenhardt KM, Furr NR, Warglien M, Sartoris C, et al. Ant colonies: Building

complex organizations with minuscule brains and no leaders. Journal of Organization Design. 2021;**10**(1):55-74

[50] Dorigo M, Bonabeau E, Theraulaz G. Ant algorithms and stigmergy. Future Generation Computer Systems. 2000; **16**(8):851-871

[51] Fazio E. A road towards the photonic hardware implementation of artificial cognitive circuits. Journal of Mental Health & Clinical Psychology. 2018;**2**(5): 1-5

[52] Blum C. Ant colony optimization: Introduction and recent trends. Physics of Life Reviews. 2005;**2**(4): 353-373

[53] Hu F, Deng Y, Aghvami AH. Scalable Multi-Agent Reinforcement Learning Algorithm for Wireless Networks, 2021: arXiv preprint arXiv:2108.00506, 2021 - arxiv.org

[54] Das S, Dey A, Pal A, Roy N. Applications of artificial intelligence in machine learning: Review and prospect. International Journal of Computer Applications. Apr 2015;**115**(9):31-41

[55] Alonzo M, Moscatelli D, Bastiani L, Belardini A, Soci C, Fazio E. All-optical reinforcement learning in solitonic x-junctions. Scientific Reports. 2018;**8**(1): 1-7

[56] Arulkumaran K, Deisenroth MP, Brundage M, Bharath AA. Deep reinforcement learning: A brief survey. IEEE Signal Processing Magazine. 2017; **34**(6):26-38

[57] Sutton RS. Introduction: The challenge of reinforcement learning. In: Machine Learning. Vol. 8. Boston. Manufactured in The Netherlands: Kluwer Academic Publishers; 19928:225-227. pp. 1-3

[58] Qinghua L, Liman W, Frutos AG, Condon AE, Corn RM, Smith LM. DNA computing on surfaces. Nature. 2000; **403**(13):175-178

[59] Aaronson S. Guest column: NP-complete problems and physical reality. ACM SIGACT News. 2005;**36**(1):30-52

[60] Caulfield H, Dolev S. Why future supercomputing requires optics. Nature Photonics. 2010;**4**:261-263

[61] Oltean M. Solving the Hamiltonian path problem with a light-based computer. Natural Computing. 2008; **7**(1):57-70

[62] Shaked NT, Messika S, Dolev S, Rosen J. Optical solution for bounded NP-complete problems. Applied Optics. 2007;**46**(5):711-724

[63] Woods D, Naughton TJ. Photonic neural networks. Nature Physics. 2012; **8**(4):257-259

[64] Anderson MC, Hulbert JC. Active forgetting: Adaptation of memory by prefrontal control. Annual Review of Psychology. 2021:1-36. DOI: 10.1146/annurev-psych-072720-094140

[65] Anderson ML. Neural reuse in the organization and development of the brain. Developmental Medicine & Child Neurology. 2016;**58**:3-6

[66] Sweatt JD. Neural plasticity and behavior—Sixty years of conceptual advances. Journal of Neurochemistry. 2016;**139**:179-199

[67] Lomonaco V. Continual Learning with Deep Architectures, PhD Dissertation thesis, Alma Mater Studiorum Università di Bologna. PhD in Computer Science and Engineering. 31 Ciclo 2019. pp. 1-134. DOI 10.6092/

unibo/amsdottorato/9073. Available from: http://amsdottorato.unibo.it/view/dottorati/DOT536/

[68] Wright CD, Liu Y, Kohary KI, Aziz MM, Hicken RJ. Arithmetic and biologically-inspired computing using phase-change materials. Advanced Materials. 2011;**23**(30):3408-3413

[69] Kuzum D, Jeyasingh RG, Lee B, Wong H-SP. Nanoelectronic programmable synapses based on phase change materials for brain-inspired computing. Nano Letters. 2012;**12**(5):2179-2186

[70] Ambrogio S, Ciocchini N, Laudato M, Milo V, Pirovano A, Fantini P, et al. Unsupervised learning by spike timing dependent plasticity in phase change memory (PCM) synapses. Frontiers in Neuroscience. 2016;**10**:56

[71] Tari H, Bile A, Moratti F, Fazio E. Sigmoid type neuromorphic activation function based on saturable absorption behavior of Graphene/PMMA composite for intensity modulation of surface Plasmon Polariton signals. Plasmonics. 2022:1-8

[72] Tait AN, De Lima TF, Zhou E, Wu AX, Nahmias MA, Shastri BJ, et al. Neuromorphic photonic networks using silicon photonic weight banks. Scientific Reports. 2017;**7**(1):1-10

[73] Zarei S, Marzban M-R, Khavasi A. Integrated photonic neural network based on silicon metalines. Optics Express. 2020;**28**(24):36668-36684

[74] Bile A, Moratti F, Tari H, Fazio E. Supervised and unsupervised learning using a fully-plastic all-optical unit of artificial intelligence based on solitonic waveguides. Neural Computing and Applications. 2021;**33**(24):17071-17079

[75] Bile A, Tari H, Fazio E. Episodic memory and information recognition using Solitonic neural networks based on photorefractive plasticity. Applied Sciences. 2022;**12**(11):5585

[76] Kandel ER. In Search of Memory: The Emergence of a New Science of Mind. New York City: WW Norton & Company; 2017

[77] Bednar JA, Miikkulainen R. Pattern-generator-driven development in self-organizing models. Computational Neuroscience. 1998;**1998**:317-323

[78] Ibarz B, Casado JM, Sanjuán MA. Map-based models in neuronal dynamics. Physics Reports. 2011;**501**(1–2):1-74

[79] Ianero B, Bile A, Alonzo M, Fazio E. Stigmergic electronic gates and networks. Journal of Computational Electronics. 2021;**20**(6):2614-2621

[80] Zhou H, Zhao Y, Wang X, Gao D, Dong J, Zhang X. Self-learning photonic signal processor with an optical neural network chip. arXiv preprint arXiv:1902.07318. 2019. DOI: 10.48550/arXiv.1902.07318

[81] Bile A, Tari H, Grinde A, Frasca F, Siani AM, Fazio E. Novel model based on artificial neural networks to predict short-term temperature evolution in museum environment. Sensors. 2022;**22**(2):615

[82] Heaton J, Goodfellow I, Bengio Y, Courville A. Deep Learning. Genetic Programming and Evolvable Machines. 2018;**19**:305-307. DOI: 10.1007/s10710-017-9314-z

[83] Hastie T, Tibshirani R, Friedman J. Overview of supervised learning. In: The Elements of Statistical Learning. Berlin, Heidelberg, New York, London, Paris, Tokyo: Springer; 2009. pp. 9-41

[84] Barlow HB. Unsupervised learning. Neural Computation. 1989;**1**(3):295-311

[85] Love BC. Comparing supervised and unsupervised category learning. Psychonomic Bulletin & Review. 2002; **9**(4):829-835

[86] Chauvet M, Bassignot F, Henrot F, Devaux F, Gauthier-Manuel L, Maillotte H, et al. Fast-beam self-trapping in LiNbO 3 films by pyroelectric effect. Optics Letters. 2015; **40**(7):1258-1261

[87] Bile A, Chauvet M, Tari H, Fazio E. Supervised learning of Soliton X-junctions in lithium niobate films on insulator. In press on Optics Letters. 2022

[88] Dao Duc K, Parutto P, Chen X, Epsztein J, Konnerth A, Holcman D. Synaptic dynamics and neuronal network connectivity are reflected in the distribution of times in up states. Frontiers in Computational Neuroscience. 2015;**9**:96

[89] Terry WS. Learning and Memory: Basic Principles, Processes, and Procedures. New York: Routledge; 2017. DOI: 10.4324/9781315622781

[90] Tulving E. Elements of Episodic Memory. Oxford University Press; 1983

Chapter 7

Application of Artificial Neural Network in Solar Energy

Bin Du and Peter D. Lund

Abstract

Accurate prediction of system performance is very important for the optimal planning of solar energy systems. The latest research of artificial neural network (ANN) technology for predicting the efficiency of solar thermal systems and the performance of photovoltaic system is reported here. Application of ANN to performance assessment of solar collectors is briefly reviewed including novel all-glass straight-through evacuated tube collectors. An overview of the most recent work of ANN for combined photovoltaic/thermal panels (PV/T) and concentrating photovoltaic collectors is also provided.

Keywords: artificial neural network, solar collector, performance prediction, thermal efficiency, photovoltaic/thermal, concentrating photovoltaics

1. Introduction

The increase of population and development of world industry requires the massive use of fossil fuel [1], resulting in environmental pollution and global warming. Renewable energy is one of the effective technical methods to alleviate this phenomenon [2]. Solar Energy is the most rapidly developing and widely used renewable energy technology. At present, there are many application forms, including solar power generation [3], seawater desalination [4], heating [5], refrigeration [6], etc. For the estimation for the efficiency of solar thermal systems, experimental study and theoretical analytic simulation codes are often utilized [7, 8]. The traditional algorithms usually employed are very complex, including the solution of complicated different equations, which usually involves large resource and takes a great quantity of time to give exact solutions [8]. Moreover, traditional analysis methods are often based on simplified assumptions, as well as simplified models and solving nonlinear partial differential equations, which reduce the prediction accuracy [9–11]. ANN is a mathematical method that mimics the behavior of human brain. It has a strong ability to learn and find nonlinear relationships between input and output in systems [12], so it has the ability to realize information processing by adjusting the connection between internal nodes [12]. Unlike complex laws and mathematical routines in traditional analysis methods, ANN can learn key information patterns in multidimensional information domain [8]. Therefore, ANN technology has obvious advantages in speed, organization ability, fault tolerance and adaptability [8]. In recent years, ANN has gained more and more applications in the solar energy field, such as solar radiation prediction [13–17], photovoltaic power generation [18–20], solar drying [21], etc.

2. Background

The solar collector converts the solar radiation energy into heat and transfers it to the heat transfer fluid [7]. Recently, the application of ANN technology in the energy-engineering systems has attracted more and more attention. ANNs have been utilized by many researchers for modeling and prediction of thermal performance of various solar collectors. Delfani et al. [22] employed ANN to determine the efficiency of direct absorption solar collector with nanofluid, and investigated the influence of collector depth and length and other important parameters on its working performance and Nusselt number. Maria et al. [23] built ANN models to evaluate the efficiency of flat plate solar collectors with silver/water nanofluid, and the results are in good agreement with the experimental data. Cuma [24] and Kalogirou et al. [25] studied various methods to predict flat plate solar collectors and solar water heater with cylindrical concentrator respectively and analyzed them comparatively. It is evident that the ANN model greatly improved the prediction accuracy.

Many ANN algorithms are employed to predict the solar heating system performance. Kumar et al. [26] investigated a roughened solar air heater, focusing on the comparison of three ANNs to evaluate its exegetic efficiency of roughened solar air heater and obviously the Radial Basis Function (RBF) model has the best performance. Abdellah et al. [27] compared the advantages and disadvantages of traditional theoretical analysis (energy balance-based) method and ANN model (data based modeling methods) in determining the performance of heat pipe solar collector. According to the results, ANN was significantly superior to other traditional theoretical methods. Kumar et al. [28] utilized ANN and multiple linear regression model to evaluate heat transfer in a solar air heater with a rough absorber and compared their performance according to a number of statistical criterias. Kumar et al. [29] further contrasted ANN models with four training functions to estimate the thermal performance of uniform flow porous bed solar air heater, and the results showed that the prediction performance of the training function was better than the other three. They also analyzed the advantages and disadvantages of three ANN algorithms for thermal performance prediction of a solar air heater with unusual physical structure [30]. Liu et al. [31] proposed an evacuated solar water heater which has high collector efficiency by developing a technology based screening method. Sadeghi et al. [32] studied the factors affecting the exergy and energy efficiency of collectors and found that the usage of copper oxide/water nanofluid in a parabolic concentrator improved the thermal efficiency. Diez et al. [33] employed various methods to evaluate the outlet temperature of working medium, and concluded after comparison that the generalized regression neural network has the best prediction effect. ANN technology with above mentioned input to estimate the characteristics of flat-plate collectors has also been presented. Comparison with conventional analytical methods indicated the superiority of ANN models [34]. Budihardjo et al. [35] modeled and analyzed heat transfer and fluid flow in single evacuated tubes. Morrison et al. [36] investigated the influence of circular heat distribution on the performance of such tubes.

3. Application of ANN in an evacuated tube solar collector

At present, the most popular evacuated tube solar collector (ETC) in the market is Dewar-tube [37, 38], because it is cheap and easy to manufacture [38]. As the fluid flow in Dewar tube is driven by buoyancy [39–41], salt usually deposit at the bottom of the tube, which worsens heat transfer. But in the all-glass straight-through evacuated tube solar collector, stronger convection promotes heat transfer, reduces heat losses and improves water quality thus eliminating the main issue of salt precipitation and weak convective heat transfer inherent in traditional Dewar-tube [37, 42]. Better performance and high efficiency [36, 43–45] decrease the cost too.

3.1 Experimental set-up

The structure of all-glass straight-through evacuated tube collector is shown in **Figure 1**. Both ends of the inner (absorption tube) and outer tube (cover glass tube) are fused together. The space between the inner and outer tubes is vacuumed with pressure <0.013 Pa to reduce convective heat loss. The selective absorption coating is coated on the outer surface of the inner tube. Hence, the working temperature of the inner tube is higher than that of the outer tube, and the temperature difference leads to thermal stress. For the safe and stable operation of the evacuated tube, the outer tube is manufactured of glass with high thermal expansion coefficient, which can withstand the thermal stress. The detailed structure of the tube is illustrated in **Table 1**.

The heat transfer fluid used in the experiment is water, which flows through the all-glass straight-through evacuated tube solar collector. The collector inlet and outlet temperatures and ambient temperature were measured by thermocouples and recorded by a data logger. The water flow rate is measured by the rotameter placed at the inlet of the tube, as illustrated in **Figure 2**.

During the experiment, water flows through the evacuated tube driven by a pump, and the flow is adjusted and stabilized by a valve connected to the flow meter. The solar radiation intensity is surveyed by a special pyrometer. The experimental site is Nanjing, China. The collected actual data include solar radiation intensity, wind speed, ambient temperature, inlet and outlet water temperature and water flow rate and. The experiment was implemented from 10 a.m. to 4 p.m. every day, and the data were recorded every 30 minutes [30].

Figure 1.
(a) Overview of the all-glass straight-through evacuated tube (b) Cross-section view of the inlet of the collector tube [51].

Parameter	Value
Length, L (m)	1.8
Absorber tube diameter, D_{abs} (m)	0.047
Outer glass tube diameter, D_{gla} (m)	0.058
Thickness of glass, ΔH (m)	0.003
Thermal conductivity of absorber, k_{abs} (W/mK)	1.2
Specific heat of absorber, C_{abs} $(J/(kg \cdot K))$	980
Absorptivity of selective coating, ξ_{abs}	0.96
Transmissivity of outer glass tube, τ_{gla}	0.96

Table 1.
Parameters of an all-glass straight through evacuated tube [11].

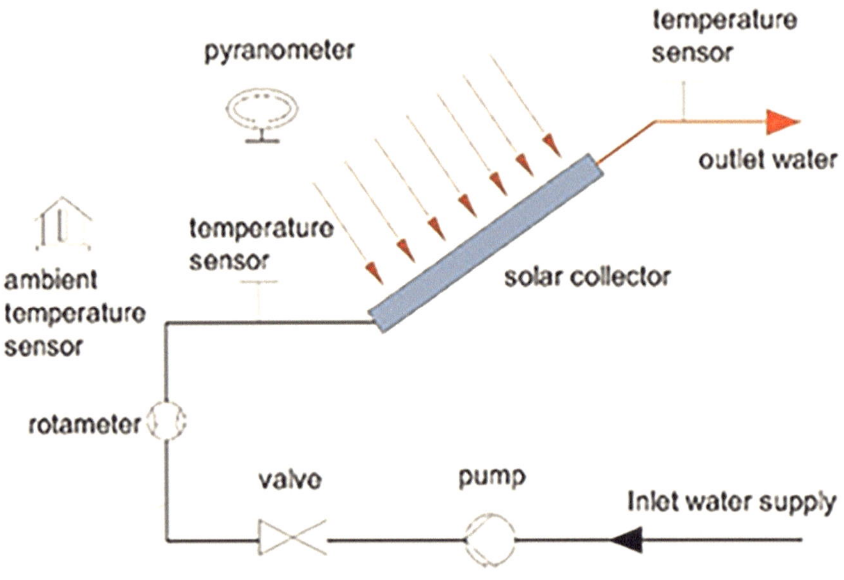

Figure 2.
Setup of the experimental system [12].

3.2 Methodology

Most of the solar energy absorbed by the evacuated tube is transferred from the tube inner wall of the tube to the working fluid flowing through the inner tube by convection heat transfer. Another part of the energy is transmitted to the inner wall of the outer glass tube through radiation and convection, and passes through the outer glass tube through heat conduction. Heat is lost to the environment by convection and to the sky by radiation from the outer surface of the outer glass tube.

Thermal efficiency is the most important criteria to evaluate the performance of evacuated tube solar collector. Here, the thermal efficiency is defined as the ratio of the heat obtained by the heat transfer fluid to the incident solar flux on the tube [30], and is written as follows [22, 46–48].

$$\eta_{th} = \frac{\dot{m}_f C_p (T_{fo} - T_{fi})}{IA_p} \tag{1}$$

where IA_p represents the solar radiation on the tube surface, T_{fi} and T_{fo} are the inlet and outlet temperature of water, respectively, $\dot{m}_f$ means the mass flow rate of heat transfer fluid, C_p is the specific heat of heat transfer fluid $(\mathbf{J}/(\mathbf{kg} \cdot \mathbf{K}))$.

3.3 ANN modeling

The Multiple linear regression (MLR), Support vector regression (SVR), Back Propagation neural network (BP) and Radial basis function (RBF) are employed here for thermal efficiency prediction of the all-glass straight-through evacuated tube collector. The following variables are used as input parameters of the models: water flow rate m_f, inlet water temperature T_{fi}, wind speed w_a, ambient temperature T_a and solar radiation intensity I. The thermal efficiency of the solar collector η_{th} is the output lay. In this work, 70% of the total 158 experimental datasets are regarded as training dataset and the other 30% is test sets. In ANN models, the optimum number of neurons in hidden layer is evaluated by the equation which is recommended by Ghritlahre et al. [7]:

$$H_n = \frac{M + N}{2} + \sqrt{T_n} \# \tag{2}$$

where H_n represents the number of hidden neurons, M and N are the input and output neurons, T_n is the number of training data.

3.3.1 Data preparation

There is likely to be a large dimension and dimension units difference between the measured data, which will seriously affect the prediction performance. Data normalization is essential to eliminate the dimensional influence among the indices. Here, the normalization is expressed as:

$$Y_{norm} = \frac{Y_i - \mathbf{mean}}{\mathbf{std}} \tag{3}$$

where mean represents the mean of the training samples, std. means the standard deviation of the training samples. The normalized data is distributed in a reasonable range, which is beneficial for further processing and analysis.

3.3.2 Performance evaluation criteria

Several criteria can be utilized to assess the accuracy of the proposed models. Their definitions are as follows:

Coefficient of Determination:

$$R^2 = 1 - \frac{\sum_{i=1}^{n} (X_{A,i} - X_{P,i})^2}{\sum_{i=1}^{n} X_{P,i}^2} \tag{4}$$

Root Mean Squared Error:

$$\mathbf{RMSE} = \sqrt{\frac{\mathbf{1}}{\mathbf{n}}\sum_{\mathbf{i=1}}^{\mathbf{n}} (\mathbf{X_{A,i}} - \mathbf{X_{P,i}})^2\#} \tag{5}$$

Mean Absolute Error:

$$\mathbf{MAE} = \frac{\mathbf{1}}{\mathbf{n}}\sum_{\mathbf{i=1}}^{\mathbf{n}} (\mathbf{X_{A,i}} - \mathbf{X_{P,i}}) \tag{6}$$

where n represents the total number of data, $X_{A,i}$ is the actual efficiency of the collector, and $X_{P,i}$ means the predicted efficiency value.

3.4 Results and discussion

After evaluation with Eq. (2), selecting 10–16 neurons in hidden layers to verify with BP algorithm, and it is obvious that the model with 13 neurons is the best.

Table 2 illustrates the comparison of RBF, BP, MLR and SVR models in predicting the thermal efficiency of all-glass straight-through evacuated tube.

It is evident that the accuracy of RBF is superior to the other methods followed by the BP model, but obviously the SVR, BP and RBF can all successfully carry out the prediction. Dealing with nonlinear problems is not the strength of MLR algorithm [49]. For nonlinear problems, SVR finds a nonlinear mapping to map the input data to the high-dimension feature space first, so that the separation status is greatly improved. Then, to classify in such feature space, and after that return to the original space, and then get the nonlinear classification of the original input space. However, after all, SVR uses linear algorithm for nonlinear regression in high-dimensional attribute space. Comparatively speaking, the major benefit of neural network method is that it is good at solving complex nonlinear relationship among variables efficiently. Thus, the deviation between the neural network model prediction values of the evacuated tube thermal efficiency and the actual data is the minimum.

The comparison between the actual data and the prediction results of the proposed models is shown in **Figure 3**. It is evident that the results of RBF model are the closest to the actual data among the four models investigated.

3.4.1 Sensitivity analysis

Sensitivity analysis refers to finding the sensitive factors that have a significant impact on the output of the models from many uncertain factors, and analyzing and calculating their impact and relative importance on the results. In short, sensitivity

Model	MAE	RMSE	R^2
MLR	0.0095	0.0121	0.6111
SVR	0.0056	0.0092	0.8447
BP	0.0053	0.0080	0.9059
RBF	0.0043	0.0066	0.9658

Table 2.
Accuracy of the models in performance prediction [11].

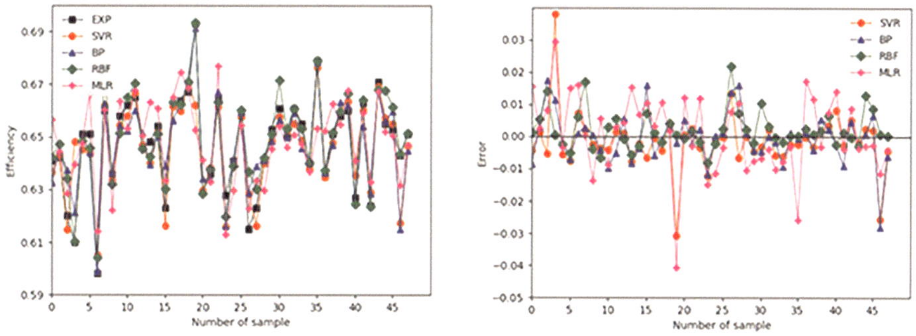

(a) (b)

Figure 3.
(a) Comparison of experimental and MLR, SVR, BP and RBF predicted thermal efficiency. (b) Individual error with MLR, SVR, BP and RBF models [11].

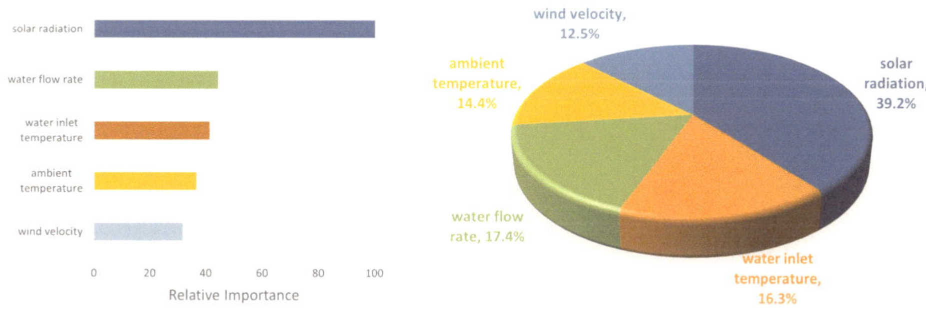

(a) (b)

Figure 4.
(a) Relative importance of input variables based on solar radiation. (b) Relative importance (%) of the inlet variables on the thermal efficiency of the evacuated tube [11].

analysis is to see which variable changes the conclusion is sensitive to [50]. Taking the RBF model in this analysis as an example, the relative importance of every input parameter to the output result is illustrated in **Figure 4**. Clearly, the solar radiation has the largest impact on the efficiency prediction of the proposed evacuated tube, followed by collector inlet temperature and water flow rate.

The efficiency value calculated by RBF model is illustrated in **Figure 5**. With the enhancement of solar radiation and the increase of water flowrate, the convective heat transfer in the tube is promoted, and the thermal performance of the evacuated tube rises. **Figure 5(b)** shows the change of thermal efficiency of evacuated tube with water flow rate and wind speed when the solar radiation intensity is 900 $\mathbf{W/m^2}$. It is visible that the increase of wind speed promotes the heat dissipation from the surface of the outer glass tube to the environment, resulting in a rise in the heat loss of the evacuated tube and a decrease in its thermal efficiency.

3.5 Combining CFD and ANN techniques modeling

The dominant energy equations of the studied all-glass straight-through evacuated tube solar collector, as well as necessary heat and mass transfer and other related

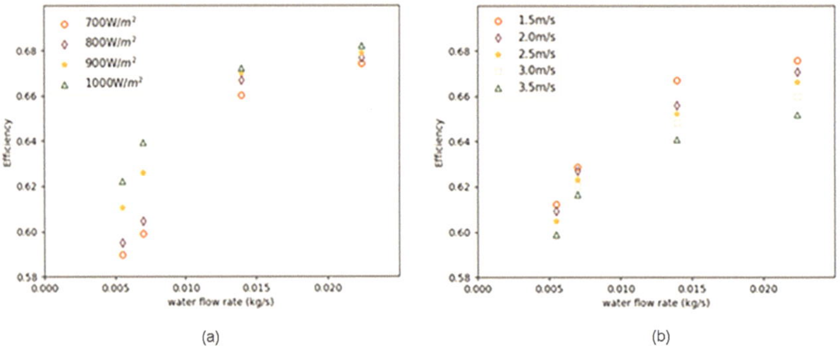

Figure 5.
(a) Efficiency vs. water flow rate (wind speed 1.5 m/s). (b) Efficiency vs. water flow rate (solar intensity 900 W/m²*) [11].*

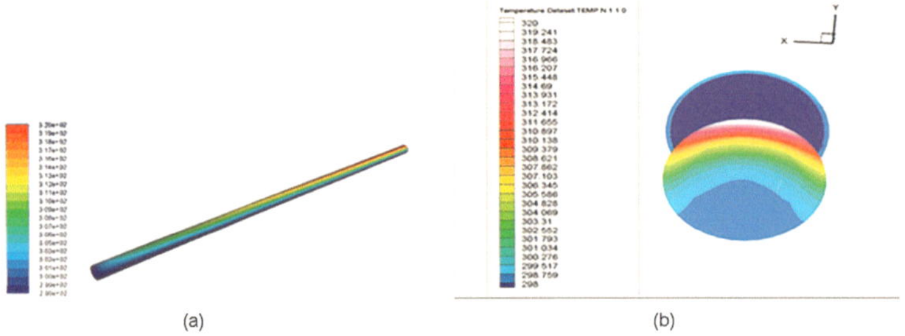

Figure 6.
(a) Temperature distribution alongside the tube. (b) Inlet and outlet temperature. Irradiance is 1000 W/m² *[51].*

conditions required for theoretical analysis have been explained in detail in [51]. The 3-D model based on these equations and conditions of the proposed evacuated tube is developed into the computational fluid dynamics (CFD) software ANSYS Fluent [47, 52, 53] to carry out the heat transfer simulation.

Figure 6 shows the temperature distribution of the evacuated tube obtained by numerical simulation of the tube model. In **Figure 6**, the inlet temperature is 298 K, the mass flow rate is 25 kg/h, and the solar radiation intensity is 1000 W/m^2.

The MLR, BP and convolutional neural network (CNN) [54, 55] models were employed to determine the thermal characteristics of all-glass straight-through evacuated tube solar collector. A total of 243 experimental data sets were employed, of which 70% were used for training and 30% were test datasets. Collector inlet water temperature, wind speed, water flow rate, ambient temperature and solar radiation intensity and the values calculated by the theoretical CFD models were used as input, the collector outlet water temperature and the thermal efficiency of the tube were regarded as output (see **Figure 7**). The output values with and without the theoretical model + CFD were compared with experimental data.

The prediction accuracies of studied models are illustrated in **Table 3**. The measurement criteria of CNN model with modeled value as one of the input parameters (CFD-CNN) is the best. Comparing the data in **Table 3**, when the modeled value of the collector outlet temperature is taken as one of the inputs, the prediction accuracies of MLR, BP and CNN models are significantly enhanced (**Figure 8**).

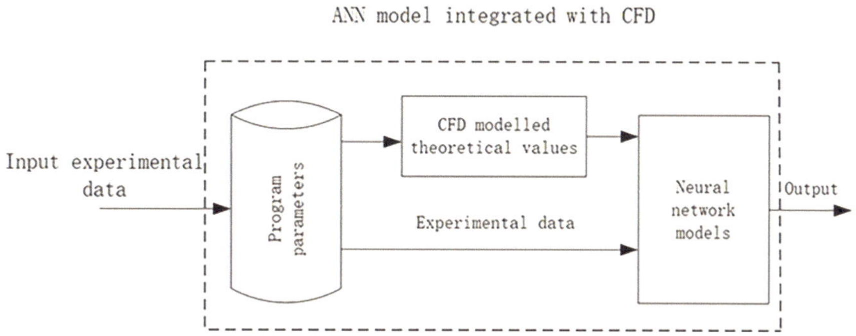

Figure 7.
Illustration of the integrated models [51].

Models	R^2		RMSE		MAE	
	Water outlet temperature	Thermal efficiency	Water outlet temperature (°C)	Thermal efficiency	Water outlet temperature (°C)	Thermal efficiency
CFD-CNN	0.9971	0.9684	0.0823	0.0044	0.0559	0.0028
CNN	0.9629	0.9548	0.3002	0.0051	0.1693	0.0036
CFD-BP	0.9937	0.9434	0.1209	0.0055	0.0910	0.0038
BP	0.9555	0.9192	0.3305	0.0067	0.2219	0.0043
CFD-MLR	0.9924	0.7443	0.0975	0.0108	0.0564	0.0086
MLR	0.7210	0.6736	0.8436	0.0112	0.6920	0.0080

Table 3.
Inaccuracies of the different models [51].

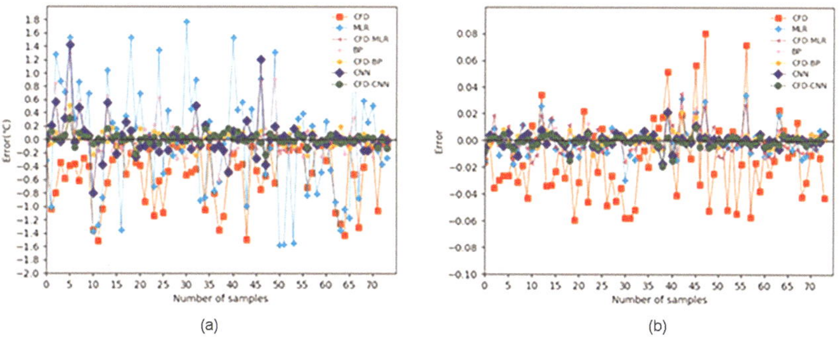

Figure 8.
Prediction error of collector (a) outlet temperature and (b) thermal efficiency by the models [51].

4. Application of ANN technique to photovoltaics

Several previous studies have reviewed the application of ANN into solar irradiance and photovoltaics (PV) power production forecasting, anomaly detection (fault diagnostic) in PV, tracking the maximum power point (MPP), etc. Here an overview of the most recent work on ANN for photovoltaic/thermal (PV/T) systems and concentrating PV(CPV) is presented.

Ammar et al. [56] investigated a PV/T based water pumping and heating system in which the PV/T panel simultaneously delivers electrical power P and thermal power Q. An ANN model was developed to determine the optimal power point (OPOP), which is defined as the crossing point of the max(P × O) curves. The focus was to calculate the optimal value of water flow with varying ambient temperature and solar radiation conditions to ensure maximum electrical power and thermal power output. The proposed neural network model regarded the solar radiation intensity and ambient temperature as the input layer and the output is the corresponding optimal water flow rate. The normal mean bias error (NMBE) was used to measure the accuracy of the ANN when the ambient temperature was 5–35°C and the solar radiation varied from 350 to 950 W/m^2, yielding a NMBE of −13.05%. The collected data was divided into cold and hot season according to the weather, for which the OPOP was computed respectively. The results show that during the hot season, with relatively stable weather conditions, the accuracy of the estimated value of the neural network model is better. The ANN algorithm provides a feasible control strategy for similar PV/T systems.

AI-Waeli et al. [57] studied a photovoltaic/thermal system using a special experimental rig for ANN analysis., Three cooling strategies were employed to verify the effectiveness of the design: PV/T with water-filled container and water as working fluid, PV/T with PCM-filled container and water as working fluid, PV/T with container filled with nanoparticles dispersed in Phase Change Material (PCM) and nanofluid as working fluid, as well as the conventional PV panel as reference. The nano-PCM and nanofluid using SiC nanoparticles yielded the best cooling effect among these methods, and the maximum efficiency reached 13.3%, while the efficiency of conventional PV was 8.1% only. Three ANN models, namely, MLP, SOFM and SVM, were used to evaluate the performance of the investigated PV/T system showing slight differences in the performance prediction.

Ahmadi et al. [58] developed ANN models such as multilayer perception (MLP), RBF, least squares support vector machine (LSSVM) and adaptive neuro-fuzzy inference system (ANFIS) to model the efficiency of a PV/T plate which contains a full circle tube as the fluid channel that is bonded to the absorber plate by special adhesives. Solar heat, solar radiation, heat, flow rate, inlet temperature were regarded as inputs, and the electrical efficiency as the output of these models. By comparing the RMSE and correlation coefficient (R^2), the LSSVM approach gave the best accuracy with R^2 = 0.9867. Using a sensitivity analysis, it was found that the inlet temperature had the greatest impact on the efficiency of the proposed PV/T system.

ANN models were also used to predict the thermal efficiency of a PV/T system that has a serpentine tube connected to the plate and using water as the cooling fluid [59]. MLP-ANN, ANFIS and LSSVM were employed to specify the thermal efficiency of the solar collector as output and inlet temperature, water flow rate and solar irradiance as input layer. The ANN model provided that best prediction performance when using the mean squared error (MSE) and determination coefficient (R^2) for the comparison. Also, here the inlet temperature proved to have the greatest impact on the thermal efficiency of the PV/T panel.

Cao et al. [60] explored six AI models, including least-squares support vector regression (LS-SVR), adaptive neuro-fuzzy inference systems (ANFIS), and four ANN methods, i.e., multi-layer perceptron (MLP), cascade feedforward (CFF), radial basis function (RBF) and generalized regression (GR) for evaluating the electrical efficiency of a PV/T system cooled by the nanofluids. Through comprehensive comparison of statistical indices such as the absolute average relative deviation (AARD), mean square error (MSE) and coefficients of determination (R^2), it was found that the ANFIS model had the best prediction accuracy for the electrical efficiency of studied PV/T system. The theoretical analysis also showed that the SiC water nanofluid was the best coolant for the PV/T system.

In [61], three ANN methods, including the radial-basis function artificial neural network (RBFANN), were employed to predict the performance of a photovoltaic thermal nanofluid (PVT/N) based collector system which is equipped with a copper sheet and tube collector and zinc-oxide (ZnO)/water nanofluid as coolant. Ten days experimental data in various weather conditions were used for training and to test of the proposed AI approach. Ambient temperature, incident solar radiation and fluid inlet temperature were regarded as input while fluid outlet temperature and electrical efficiency were set as the output layer. The ANFIS was more accurate for predicting the fluid outlet temperature, but the RBFANN was superior to the other methods to predict the electrical efficiency of the proposed PVT/N unit.

Renno et al. [62] compared the prediction performance of Random-Forest (RF), ANN and Linear Regression Model (LRM) approaches to predict the temperature of multi-junction solar cells. The studied cells constituted of InGaP/GaAs/Ge and InGaP/InGaAs/Ge under a high concentration Fresnel lens. The input variables were the local hour, global radiation, concentration factor and the environmental temperature，and the cell temperature was used as output. The RF method yielded the best performance with the lowest values of RMSE, MAE and MAPE. It was is observed that the cell temperature increased with the increasing ambient temperature, solar radiation, and concentration ratio.

In [63], the power output of a V-trough photovoltaic system was predicted with support vector machine (SVM), ANN, kernel and nearest-neighbor and deep learning (DL) methods. Through a statistical indices comparison, the support vector machine gave better performance prediction accuracy, although all the presented algorithms predicted the PV module power output satisfactorily. Also, the ANN model was not inferior to the SVM algorithm in evaluating the peak data. The prediction performance of DL and ANN were also compared with SVM. The results showed that the predicted PV power output by DL was higher than the actual data, which was likely due to the availability of data.

5. Conclusions

Application of artificial neural network for performance prediction of solar energy collectors has briefly been introduced here including comparison to traditional analysis methods.

Back propagation (BP), radial basis function (RBF), support vector regression (SVR) and multiple linear regression (MLR) were used to predict the performance of a novel all-glass straight-through evacuated tube solar collector employing experimental datasets. The RBF and BP outperformed the SVR and MLR methods, but the accuracy of the first three models mentioned above were well within acceptable limits (R^2s were 0.8447, 0.9059 and 0.9658, respectively). However, the MLR algorithm was not good in dealing with nonlinear problems. The RBF method showed the best performance with the lowest RMSE (0.0066) and the lowest MAE (0.0043) for the solar collector efficiency prediction.

A novel approach combining mathematical performance simulation (CFD) and neural networks was also investigated for determining the performance of the all-glass straight-through evacuated tube. The results show that regarding CFD modeled output as the input of ANN significantly improved the evaluation accuracy of all proposed models including MLR, BP and convolutional neural network (CNN). The CFD-CNN model is superior to the other studied models with the highest R^2 and the lowest RMSE, 0.9684 and 0.0044 (**Table 3**).

The research on applying ANN to photovoltaics was also reported with focus on the utilization of neural networks for output power prediction of photovoltaic/thermal system (PV/T) and concentrating photovoltaics (CPV). The review demonstrated the usefulness of ANN also for the PV field.

Future work of ANN in solar energy could extend to other design parameters and meteorological data as input to the neural network model. Also, using new ANN approaches such as the recurrent neural network could be relevant. Future directions of interest include the combination of some metaheuristic methods such as gray wolf optimization (GWO), genetic algorithm (GA), particle swarm optimization (PSO) and ANN to optimize ANN structure and improve ANN performance. Extensions of ANN, e.g., extreme machine learning (EML), adaptive network-based fuzzy inference system can be used to improve prediction accuracy. Based on the work presented here, it is believed that the artificial neural network will increasingly be applied in the field of solar energy.

Acknowledgements

Part of this work was funded by the National Natural Science Foundation of China (Grant number 51736006). The support of Aalto University is also acknowledgement.

Author details

Bin Du[1,2]* and Peter D. Lund[1,3]

1 Key Laboratory of Solar Energy Science and Technology in Jiangsu Province, Institue of Energy and Environment, Southeast University, Nanjing, China

2 Energy Storage Research Center, Southeast University, Nanjing, China

3 School of Science, Aalto University, Espoo, Finland

*Address all correspondence to: bindubill@hotmail.com

References

[1] Can S, Sharp Julia L, Annick AA. Factors impacting diverging paths of renewable energy: A review. Renewable Sustainable Energy Reviews. 2018;**81**: 2335-2342

[2] Guven G, Sulun Y. Pre-service teacher's knowledge and awareness about renewable energy. Renewable Sustainable Energy Reviews. 2017;**80**: 663-668

[3] Jain S, Kumar Jain N, Jamie VW. Challenges in meeting all of India's electricity from solar: An energetic approach. Renewable Sustainable Energy Reviews. 2018;**82**:1006-1013

[4] Chen L, Huiyao W, Sarada K, Krishna K, Xu P. Low-cost and reusable carbon black based solar evaporator for effective water desalination. Desalination. 2020;**483**:1-15

[5] Pereira Da Cunha J, Eames PC. Compact latent heat storage decarbonization potential for domestic hot water and space heating applications in the UK. Applied Thermal Engineering. 2018;**134**:396-406

[6] Salilih Elias M, Birhane YT. Modelling and performance analysis of directly coupled vapor compression solar refrigeration system. Solar Energy. 2019; **190**:228-238

[7] Ghritlahre HK, Prasad RK. Application of ANN technique to predict the performance of solar collector system—A review. Renewable Sustainable Energy Reviews. 2018;**84**(3): 75-88

[8] Kalogirou SA. Applications of artificial neural-networks for energy systems. Applied Energy. 2000; **67**:17-35

[9] Elsheikh AH, Sharshir SW, Elaziz MA, Kabeel AE, Wang GL, Zhang H. Modeling of solar energy systems using artificial neural network: A comprehensive review. Solar Energy. 2019;**180**:622-639

[10] Bellos E, Tzivanidis C. Development of an analytical model for the daily performance of solar thermal systems with experimental validation. Sustainable Energy Technology Assessments. 2018;**28**:22-29

[11] Du B, Lund PD, Wang J, Kolhe M, Eric H. Comparative study of modelling the thermal efficiency of a novel straight through evacuated tube collector with MLR, SVR, BP and RBF methods. Sustainable Energy Technologies and Assessments. 2021;**44**:1-10

[12] Du B, Peter D, Lund WJ. Improving the accuracy of predicting the performance of solar collectors through clustering analysis with artificial neural network models. Energy Reports. 2022;**8**: 3970-3981

[13] Vakili M, Sabbagh-Yazdi SR, Khosrojerdi S, Kalhor K. Evaluating the effect of particular matter pollution on estimation of daily global solar radiation using artificial neural network modeling based on meteorological data. Journal of Cleaner Production. 2017;**141**:1275-1285

[14] Mghouchi YE, Chham E, Zemmouri EM, Bouardi AEI. Assessment of different aombinations of meteorological parameters for predicting daily global solar radiation using artificial neural network. Building and Environment. 2019;**149**:607-622

[15] Bou-Rabee M, Sulaiman SA, Saleh MS, Marafi S. Using artificial neural networks to estimate solar

radiation in Kuwait. Renewable and Sustainable Energy Reviews. 2017;**72**: 434-438

[16] Shaddel M, Javan DS, Baghernia P. Estimation of hourly global solar irradiation on tilted absorbers from horizontal one using artificial neural network for case study of Mashhad. Renewable and Sustainable Energy Reviews. 2016;**53**:59-67

[17] Kashyap Y, Bansal A, Sao AK. Solar radiation forecasting with multiple parameters neural networks. Renewable and Sustainable Energy Reviews. 2015; **49**:825-835

[18] Hussain M, Dhimish M, Titarenko S, Mather P. Artificial neural network based photovoltaic fault detection algorithm integrating two bi-directional input parameters. Renewable Energy. 2020;**155**:1272-1292

[19] Yadav AK, Sharma V, Malik H, Ghandel SS. Daily array yield prediction of grid-interactive photovoltaic plant using relief attribute evaluator based radial basis function neural network. Renewable and Sustainable Energy Reviews. 2018;**81**:2115-2127

[20] Almonacid F, Fernandez EF, Mellit A, Kalogirou S. Review of techniques based on artificial neural networks for the electrical characterization of concentrator photovoltaic technology. Renewable and Sustainable Energy Reviews. 2017;**75**: 938-953

[21] Prakash O, Laguri V, Pandey A, Kumar A. Review on various modelling techniques for the solar dryers. Renewable and Sustainable Energy Reviews. 2016;**62**:396-417

[22] Shahram D, Mostafa E, Maryam K. Application of artificial neural network for performance prediction of a nanofluid-based direct absorption solar collector. Sustainable Energy Technologies and Assessments. 2019;**36**(12):1-11

[23] Maria TA, Nizar A, Subathra MSP, Godson AL. Analysing the performance of a flat plate solar collector with silver/ water nanofluid using artificial neural network. Procedia Computer Science. 2016;**93**:33-40

[24] Cuma C, Fethi H, Hamit C, Imdat T. Generating hot water by solar energy and application of neural network. Applied Thermal Engineering. 2005;**25**: 1337-1348

[25] Kalogirou SA. Prediction of flat-plate collector performance parameters using artificial neural networks. Solar Energy. 2006;**80**:248-259

[26] Kumar GH, Krishna PR. Exegetic performance prediction of solar air heater using MLP, GRNN and RBF models of artificial neural network technique. Journal of Environmental Management. 2018;**223**:566-575

[27] Abdellah S, Hossein P, Mehdi K. Comparative and performative investigation of various data-based and conventional theoretical methods for modelling heat pipe solar collectors. Solar Energy. 2020;**198**:212-223

[28] Kumar GH, Krishna PR. Prediction of heat transfer of two different types of roughened solar air heater using artificial neural network technique. Thermal Science and Engineering Progress. 2018;**8**: 145-153

[29] Kumar GH, Krishna PR. Prediction of thermal performance of unidirectional flow porous bed solar air heater with optimal training function using artificial neural network. Energy Procedia. 2017; **109**:369-376

[30] Kumar GH, Krishna PR. Investigation of thermal performance of unidirectional flow porous bed solar air heater using MLP, GRNN and RBF models of ANN technique. Thermal Science and Engineering Progress. 2018; **6**(6):226-235

[31] Zhj L, Hao L, Liu Kj Y, hch, Cheng Kw. Design of high-performance water-in-glass evacuated tube solar water heaters by a high-throughput screening based on machine learning: A combined modelling and experimental study. Solar Energy. 2017;**142**:61-67

[32] Sadeghi G, Nazari S, Ameri M, Shama F. Energy and exergy evaluation of the evacuated tube collector using Cu_2O/water nanofluid utilizing ANN methods. Sustainable Energy Technologies and Assessments. 2020;**37**: 1-14

[33] Diez FJ, Navas-Gracia LM, Martinez-Rodriguez A, Correa-Guimaraes A, Chico-Santamarta L. Modelling of a flat-plate solar collector using artificial neural networks for different working fluid(water) flow rates. Solar Energy. 2019;**188**: 1320-1331

[34] Sozenm A, Menlikm T, Unvar S. Determination of efficiency of flat-plate solar collectors using neural network approach. Expert Systems with Applications. 2008;**35**:1553-1539

[35] Budihardjo I, Morrison GL, Behnia M. Measurement and simulation of flow rate in a water-in-glass evacuated tube solar collectors. Solar Energy. 2007; **81**:1460-1472

[36] Morrison GL, Budihardjo I, Behnia M. Measurement and simulation of flow rate in a water-in-glass evacuated tube solar water heater. Solar Energy. 2005;**78**(2):257-267

[37] Kim Y, Seo T. Thermal performances comparisons of the glass evacuated tube collectors with shapes of absorber tube. Renewable Energy. 2007;**32**: 772-795

[38] Shf Q, Ruth M, Ghosh S. Evacuated tube collectors: A notable driver behind the solar water heater industry in China. Renewable and Sustainable Energy Reviews. 2015;**47**:580-588

[39] Daghigh R, Shafieian A. Theoretical and experimental analysis of thermal performance of a solar water heating system with evacuated tube heat pipe collector. Applied Thermal Engineering. 2016;**103**:1219-1227

[40] Gao Y, Zhang Q, Fan R, Liu X, Yu Y. Effects of thermal mass and flow rate on forced-circulation solar hot-water system: Comparison of water-in-glas and U-pipe evacauetd-tube solar collectors. Solar Energy. 2013;**98**:290-301

[41] Ayompe LM, Duffy A. Thermal performance analysis of a solar water heating system with heat pipe evacuated tube collector using data from a field trial. Solar Energy. 2013;**90**:17-28

[42] Salgado-Conrado L, Lopez-Montelongo A. Barriers and solutions of solar water heaters in Mexican household. Solar Energy;**188**:831-838

[43] Li JR, Li XD, Wang Y, Tu JY. A theoretical model of natural circulation flow and heat transfer within horizontal evacuated tube considering the secondary flow. Renewable Energy. 2020;**147**(3):630-638

[44] Sobhansarbandi S, Martinez PM, Papadimitratos A, Zakhidov A, Hassanopour F. Evacuaetd tube solar collector with multifunctional absorber layers. Solar Energy. 2017;**146**(4): 342-350

[45] Budihardjo I, Morrison GL, Behnia M. Natural circulation flow through water-in-glass evacuated tube solar collectors. Solar Energy;**81**(12): 1460-1472

[46] Esen H, Ozgen F, Esen M, Sengur A. Artificial neural network and wavelet neural network approached for modelling of a solar air heater. Expert Syatems with Applications. 2009;**36**(10): 11240-11248

[47] Tagliafico LA, Scarpa F, Rosa MD. Dynamic thermal models and CFD analysis for flat-plate thermal solar collectors-a review. Renewable and Sustainable Energy Reviews. 2014;**30**(2): 526-537

[48] Shafieian A, Osman JJ, Khiadani M, Nosrati A. Enhancing heat pipe solar water heating systems performance using a novel variable mass flow rate technique and different solar working fluids. Solar Energy. 2019;**186**(5): 191-203

[49] Khatib T, Mohamed A, Sopian K. A review of solar energy modelling techniques. Renewable and Sustainable Energy Reviews. 2012;**16**(6):2864-2869

[50] Alvarez ME, Hernandez JA, Bourouis M. Modelling the performance parameters of a horizontal falling film absorber with aqueous (lithium, potassium, sodium) nitrate solution using artificial neural networks. Energy. 2016;**102**(5):313-323

[51] Bin D, Lund PD, Wang J. Combining CFD and artificial neural network techniques to predict the thermal performance of all-glass straight evacuated tube solar collector. Energy. 2021;**220**:1-15

[52] Filipovic P, Dovic D, Ranilovic B, Horvat I. Numerical and experimental approach for evaluation of thermal performance of a polymer solar collector. Renewable and Sustainable Energy Reviews. 2019;**112**(9):127-139

[53] Alfaro-Ayala JA, Martinez-Rodriguez G, Picon-Nunez M, Uribe-Ramirez AR, Gallegos-Munoz A. Numerical study of a low temperature water-in-glass evacuated tube solar collector. Energy Conversion and Management. 2015; **94**(4):472-481

[54] Ahmed R, Sreeram V, Mishra Y, Arif MD. A review and evaluation of the state-of-the-art in PV solar power forecasting: Techniques and optimization. Renewable and Sustainable Energy Reviews. 2020;**124**:1-26. Article 109792

[55] Feng C, Zhang J, SolarNet. A sky image-based deep convolutional neural network for intra-hour solar forecasting. Solar Energy. 2020;**204**(7):71-78

[56] Ammar MB, Chaabene M, Chtourou Z. Artificial neural network based control for PV/T panel to track optimum thermal and electrical power. Energy Conversion and Management. 2013;**65**(1):372-380

[57] Ali HA, Al-Waeli KS, Kazem HA, Yousif JH, Miqdam TC, Ibrahim A, et al. Comparison of prediction methods of PV/T nanofluid and nano-PCM system using a measured dataset and artificial neural network. Solar Energy. 2018; **162**(3):378-396

[58] Ahmadi MH, Baghban A, Sadeghzadeh M, Zamen M, Mosavi A, Shamshirband S, et al. Evaluation of electrical efficiency of photovoltaic thermal solar collector. Engineering Applications of Computational Fluid Mechanics. 2020;**14**(1):545-565. DOI: 10.1080/19942060.2020.1734094

[59] Zamen M, Alireza Baghban S, Pourkiaei M, Ahmadi MH. Optimization methods using artificial intelligence algorithms to estimate thermal efficiency of PV/T system. Energy Science and Engineering. 2019;**7**(2):821-824. DOI: 10.1002/ese3.312

[60] Cao Y, Kamrani E, Mirzaei S, Khandakar A, Vaferi B. Electrical efficiency of the photovoltaic/thermal collectors cooled by nanofluids: Machine learning simulation and optimization by evolutionary algorithm. Energy Reports. 2022;**8**(1):24-36

[61] Kalani H, Sardarabadi M, Passandideh-Frad M. Using artificial neural network models and particle swarm optimization for manner prediction of a photovoltaic thermal nanofluid based collector. Applied Thermal Engineering. 2017;**113**(2): 1170-1177

[62] Renno C, Petito F. Triple-junction cell temperature evaluation in a CPV system by means of random-Forest model. Energy Conversion and Management. 2018;**169**(5):124-136

[63] Agbulut U, Gurel AE, Ergun A, Ceylan I. Performance assessment of a V0trough photovoltaic system and prediction of power output with different machine learning algorithms. Journal of Cleaner Production. 2020; **268**(122269):1-12

Chapter 8

Modeling a Petrochemical Unit with Artificial Neural Networks (ANN)

Shafaati Akbar and Pourazad Hamidreza

Abstract

The purpose of this chapter is to model a petrochemical unit by neural networks to estimate the product flow rate of the plant by it. Multilayer perceptron and RBF neural networks have been used in this work, and finally, the outputs of both types of networks have been compared to choose the more accurate network. The same data have been used for training and modeling both networks. The data used for this modeling have been collected by measuring the flow rate of input materials and output products from the plant in ton per day. **Table 1** shows the input materials and products.

Keywords: artificial neural networks, RBF, MLP, regression, petrochemical unit

1. Introduction

To model a petrochemical unit by the artificial neural network, the necessary acquaintances with artificial neural networks should be made first, and we should answer the question of why we should use artificial neural networks instead of conventional methods.

The artificial neural network is a complex nonlinear computing system that is inspired by nature, and the main advantage of this network in performing calculations compared with other computing systems is because of its internal structure [1].

Neural networks are composed of a large number of neurons that have extensive connections with each other. These neurons have the ability to share information with each other. A neural network performs calculations by organizing neurons and communication between them and the information stored in them.

Using conventional modeling methods requires a lot of mathematical calculations and has many complications, especially when we are dealing with a nonlinear system. It takes a lot of time to do this, and if there is an error in the calculations, all the steps must be repeated, and the existing error must be identified and fixed. On the other hand, all the influencing parameters of the designed model should be considered, and a relationship should be defined for how it affects the system, and finding these relationships also has complications. Finding these relationships is important because it can have a great impact on the accuracy of the designed model's output. Finally, all relevant equations must be solved, which is very time-consuming.

IntechOpen

There is no need to perform complex mathematical calculations in modeling with a neural network, and we can save time. Other advantages of neural networks compared with other methods include adaptability, nonlinearity, error tolerance, and flexibility against changing conditions.

To model with an artificial neural network, a dataset is needed to train the network, and these data must be collected by experimental tests, industrial devices, etc.

For example, in the modeling of a petrochemical unit, the goal is to predict outputs according to the flow rate of input to the petrochemical unit, so to prepare basic data for training the network, it is necessary to flow rate of inputs and outputs, during different operations be measured and entered into the network as training data.

In this modeling, the entire petrochemical unit considered as a black box (**Figure 1**), and only the flow rates of input and output materials are considered as influencing parameters. This is because none of the processes that take place inside the petrochemical unit are involved in the model designed based on neural networks.

Here are several related works:

Tufaner et al. developed a three-layer artificial neural network (ANN) and nonlinear regression model to predict the performance of biogas production from the anaerobic hybrid reactor (AHR). In this study, experimental data were used to estimate the biogas production rate with models produced using both ANNs and nonlinear regression methods. Moreover, 10 related variables, such as reactor fill ratio, influent pH, effluent pH, influent alkalinity, effluent alkalinity, organic loading rate, effluent chemical oxygen demand, effluent total suspended solids, effluent suspended solids, and effluent volatile suspended solids, were selected as inputs of the model [2].

DS Pandey et al. developed a multilayer feed-forward neural network to predict the lower heating value of gas (LHV), lower heating value of gasification products including tars and entrained char (LHVp), and syngas yield during gasification of municipal solid waste (MSW) during gasification in a fluidized bed reactor. These artificial neural networks (ANNs) with different architectures are trained using the Levenberg-Marquardt (LM) back-propagation algorithm. Nine input and three output parameters are used to train and test various neural network architectures in both multiple-output and single-output prediction paradigms using the available experimental datasets [3].

M. EI-Sefy et al. developed a feed-forward back-propagation artificial neural network (ANN) model and trained to simulate the interaction between the reactor core and the primary and secondary coolant systems in a pressurized water reactor. A Nuclear Power Plant (NPP) is a complex dynamic system of systems with highly nonlinear behaviors. In order to control the plant operation under both normal and abnormal conditions, the different systems in NPPs (e.g., the reactor core components, primary and secondary coolant systems) are usually monitored continuously, resulting in very large amounts of data. The transients used for model training

Figure 1.
Assumed structure for the petrochemical unit for modeling by artificial neural network.

included perturbations in reactivity, steam valve coefficient, reactor core inlet temperature, and steam generator inlet temperature. Uncertainties of the plant physical parameters and operating conditions were also incorporated in these transients [4].

1.1 Introduction to radial basis function networks (RBF)

Radial basis neural networks use the radial basis function instead of the logistic function as the activation function. The logistic function maps some arbitrary value to a 0–1 interval to answer a yes or no question (binary question) [5].

These types of neural networks are suitable for "classification" and "decision-making systems," but they are not good in Continuous values. While the basic radial basis function answers the question, how far are we from the goal? And this makes these neural networks suitable for function approximation and machine control (for example, as an alternative to the PID controller) [5].

Radial basis neural networks are special types of natural neural networks that are distance-based and measure the similarity between data based on distance.

Unlike MLP networks, which have multiple consecutive layers, the RBF network consists of three fixed layers. An input layer, which is the input data entered into the network from there, the middle layer, which contains radial basis functions, and the output layer, which gives a linear combination of all middle layer outputs.

Output layer uses a linear activation function or can be thought of without any activation function [6].

1.2 Introduction to multilayer perceptron networks (MLP)

One of the most basic neural models available is the multilayer perceptron model, which simulates the transfer function of the human brain. In this type of neural network, most of the network behavior of the human brain and signal propagation have been considered in it, and hence, they are sometimes called feed-forward networks [1].

Perceptron is a machine learning algorithm that is in the field of supervised learning. This algorithm is known as one of the first artificial neural network algorithms used in this field. Perceptron is considered a type of binary classification algorithm, which means that this algorithm can decide whether a member belongs to a specific category or not [7].

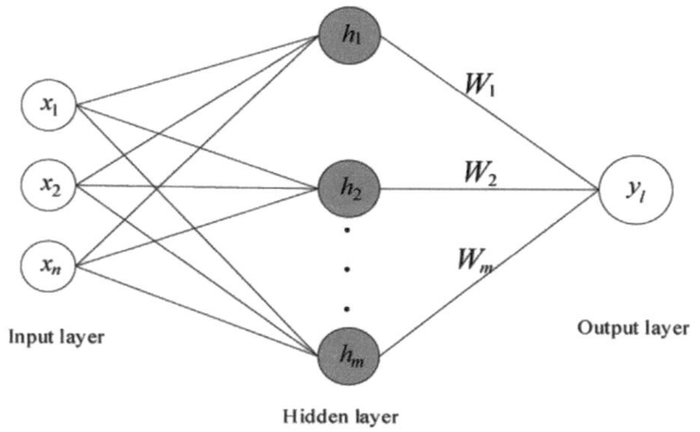

Figure 2.
Schematic of an RBF neural network.

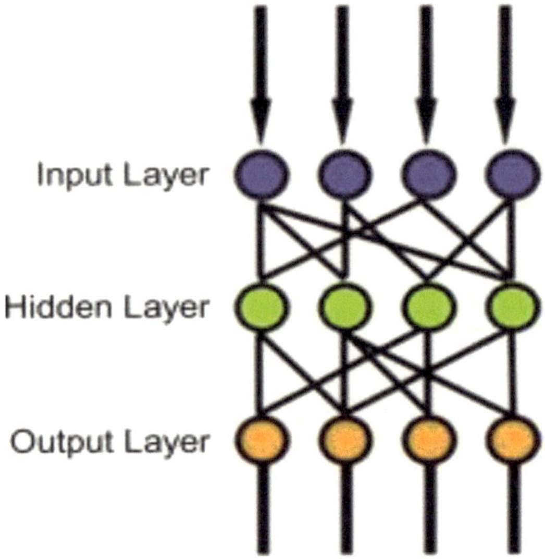

Figure 3.
Schematic of an MLP neural network.

A multilayer perceptron neural network consists of at least three layers, which are the input layer, a hidden layer, and the output layer. In this type of artificial neural network, the outputs of the first (input) layer are used as the inputs of the next (hidden) layer. This continues until, after a certain number of layers, the outputs of the last hidden layer are used as the inputs of the output layer. All the layers that are placed between the input layer and the output layer are called "Hidden Layers" (**Figures 2** and **3**).

2. Modeling by radial basis function networks (RBF) neural network

The necessary dataset for training this network is by measuring the flow rate of input materials and products (outputs) that have been collected, which has been measured every day for a year. **Table 1** shows the inputs and outputs of the petrochemical unit.

For testing the network, experimental data were given to the network and the networks outputs compared with the real outputs of petrochemical unit shown in the **Figure 4**.

The empty circles on the blue graph in **Figure 4** indicate the measured amount of the products (experimental data or targets), and the empty circles on the orange graph also indicate the predicted parameters by the neural network. Some of these circles are almost coincident with each other, and some are slightly different from each other. In the best case, these points should overlap. The names of each of which are indicated by arrows.

To better understand the amount of difference and whether the network has provided an acceptable performance or not, we can use linear regression between the data estimated by the network and the measured parameters (experimental data). **Figure 5** shows the regression between the experimental and predicted data used in **Figure 4**.

Inputs	Outputs
SRG	POLYETHYLENE
LPG	HIPS
PBR	GPPS
BENZENE	EPS
ACRYLONITRILE	ABS
BUTEN-1	TOLUENE
MINERAL OIL	BD
GAS(Nm^3/d)	PENTANE
STEAM	PROPYLENE
WATER(m^3)	C4 RAFFINATE
POWER	C.F.O
STYRENE	FUEL OIL
—	C7-C9

Table 1.
Inputs and outputs of the petrochemical unit.

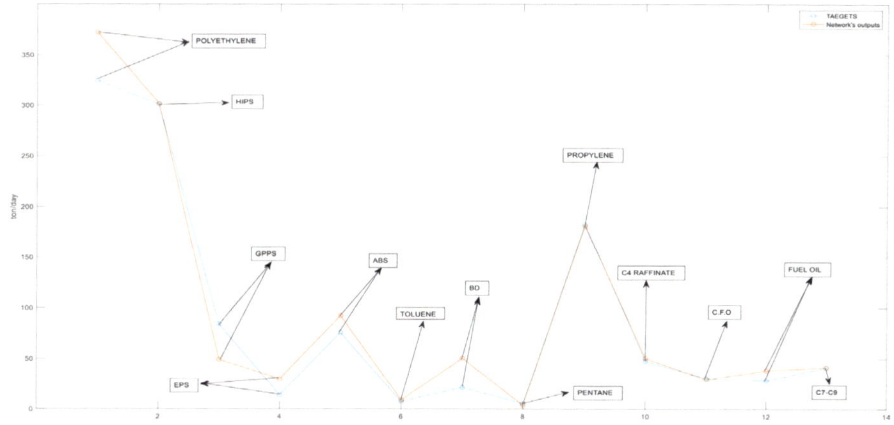

Figure 4.
Difference between RBF network's outputs and experimental data (targets).

As can be seen, the correlation coefficient between the estimated and experimental data is 0.987, which is acceptable for the petrochemical unit in non-essential and non-sensitive situations.

3. Modeling by multilayer perceptron (MLP) neural network

The multilayer perceptron network considered for this modeling consists of three layers. The first layer has 80, the second layer has 35, and the third layer has 13 neurons.

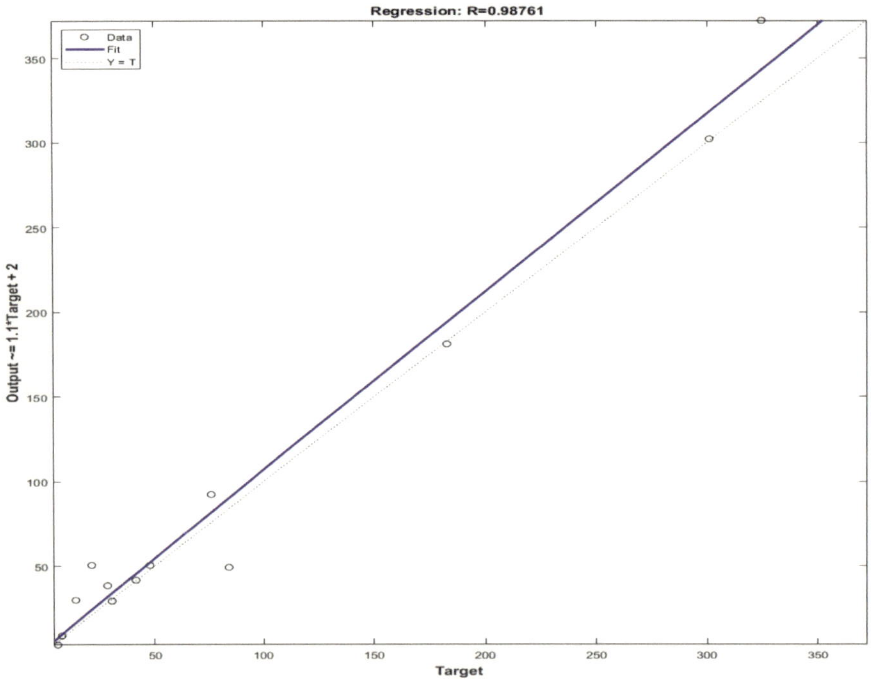

Figure 5.
Regression between predicted data with RFB network and experimental data.

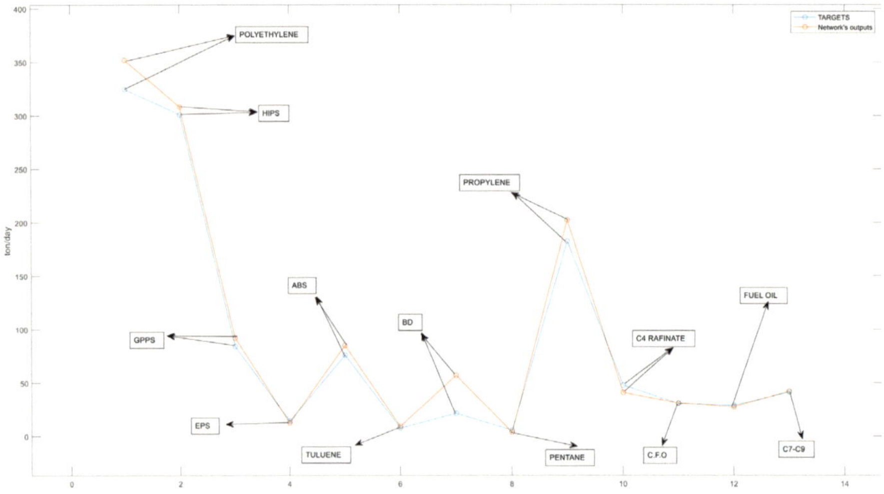

Figure 6.
Difference between multilayer perceptron network's outputs and experimental data(targets).

The activation functions considered for each of these layers are **Relu** for the first and second layers and **purlin** for the last layer, respectively.

The best performance of the network is achieved when it gives the value of the error between the network and the experimental data to the lowest possible value, and this is done by some functions, which are called performance functions. In this

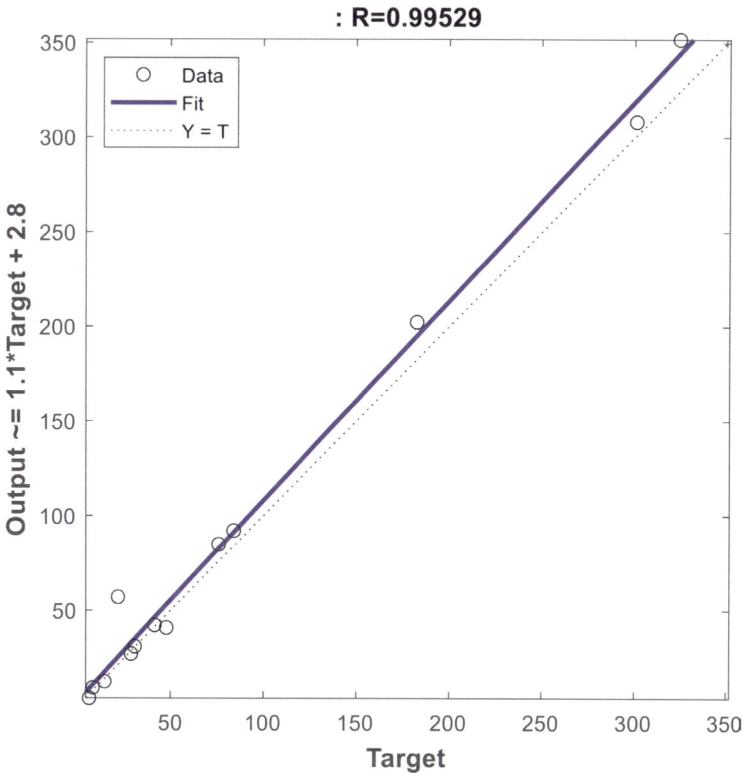

Figure 7.
Regression between predicted data with MLP network and experimental data.

modeling, the mean square error (MSE) performance function is used. For this modeling, the Levenberg Marquardt training algorithm is used [1, 8].

After the training process, the same data used for testing the RBF network in **Figure 4** are used to test the MLP network, and the results are shown in **Figure 6**.

Like **Figure 4**, the blue graph represents the experimental data, and the orange graph represents the data estimated by the neural network. The empty circles on the blue graph in **Figure 7** indicate the measured amount of the products (experimental data), and the empty circles on the orange graph also indicate the estimated parameters by the neural network.

As before, to better understand the amount of difference and whether the network has provided an acceptable performance or not, we can use linear regression between the data predicted by the network and the measured parameters (experimental data).

The correlation coefficient between experimental and estimated data by the network is equal to 0.995, which indicates the good performance of the network.

4. Conclusion

By comparing the correlation coefficient of the RBF neural network, which is equal to 0.987, and the correlation coefficient of the MLP neural network, which is equal to 0.995, it can be concluded that the MLP neural network can perform better in estimating the amount of petrochemical unit products in different conditions. Due to the complex processes that are carried out inside the petrochemical unit, by which inputs are converted into products, a large number of experimental samples are needed for modeling by neural network and its training. In other words, we should take samples from everything that has a direct or indirect effect on the system under study (petrochemical unit) and changes the amount of products produced by the petrochemical unit. It means recording the amount of these changes and finally preparing the required dataset. It is obvious that one of the things that have a great effect on the amount of products produced from a petrochemical unit is the amount of input materials (feed). Therefore, the amount of changes in production products that occur due to changes in the amount of feed should be recorded. These changes were measured per ton per day. As it was said, complex processes take place inside the Petrochemical unit, such as chemical reactors, distillation towers, etc., each of which has an effect on the amount of production, but due to the limitation in measuring these factors, it was decided to measure only the amount of input feed and changes in the amount of produced products, and for this reason, we omitted the details and processes within the petrochemical unit. By doing this, the accuracy of the designed neural network was disrupted, and to solve this problem, it was decided to increase the number of samples collected from the amount of input materials (feed) and changes in the products produced, so that the neural network has more data for training. It took a year to collect this amount of data to complete the desired dataset.

Author details

Shafaati Akbar[1*] and Pourazad Hamidreza[2]

1 Urmia University of Technology, Urmia, Iran

2 Sahand University of Technology, Tabriz, Iran

*Address all correspondence to: akbar007.78.sh@gmail.com

References

[1] Haykin SS. Neural Networks and Learning Machines/Simon Haykin. New York: Prentice Hall; 2009. ch4, ch5

[2] Tufaner F, Demirci Y. Prediction of biogas production rate from anaerobic hybrid reactor by artificial neural network and nonlinear regressions models. Clean Technologies and Environmental Policy. 2020;**22**(3):713-724

[3] Pandey DS et al. Artificial neural network based modeling approach for municipal solid waste gasification in a fluidized bed reactor. Waste Management. 2016;**58**:202-213

[4] El-Sefy M et al. Artificial neural network for predicting nuclear power plant dynamic behaviors. Nuclear Engineering and Technology. 2021;**53**(10):3275-3285

[5] The Mostly Complete Chart of Neural Networks Explained. 2021. Available from: https://towardsdatascience.com/the-mostly-complete-chart-of-neural-networks-explained-3fb6f2367464

[6] Radial Basis Function Neural Network Simplified. 2021. Available from: https://towardsdatascience.com/radial-basis-function-neural-network-simplified-6f26e3d5e04d

[7] Multi-Layer Perceptron. Available from: https://www.sciencedirect.com/topics/computer-science/multilayer-perceptron

[8] Haykin SS. Neuronal Networks: A Comprehensive Foundation. Subsequent edition. New York: Prentice Hall; 2000. ch1, ch3